INTERNATIONAL

REVIEW OF CYTOLOGY

VOLUME 91

Membranes

INTERNATIONAL

Review of Cytology

EDITED BY

G. H. BOURNE
St. George's University School of Medicine
St. George's, Grenada
West Indies

J. F. DANIELLI
Danielli Associates
Worcester, Massachusetts

ASSISTANT EDITOR

K. W. Jeon
Department of Zoology
University of Tennessee
Knoxville, Tennessee

VOLUME 91

Membranes

EDITED BY

J. F. DANIELLI
Danielli Associates
Worcester, Massachusetts

ACADEMIC PRESS, INC. 1984
(*Harcourt Brace Jovanovich, Publishers*)
Orlando   San Diego   New York   London
Toronto   Montreal   Sydney   Tokyo

ACADEMIC PRESS, INC.
Orlando, Florida 32887

United Kingdom Edition published by
ACADEMIC PRESS, INC. (LONDON) LTD.
24/28 Oval Road, London NW1 7DX

LIBRARY OF CONGRESS CATALOG CARD NUMBER: 52-5203
ISBN 0-12-364491-7

PRINTED IN THE UNITED STATES OF AMERICA

84 85 86 87   9 8 7 6 5 4 3 2 1

Contents

Supramolecular Cytology of Coated Vesicles

RICHARD E. FINE AND COLIN D. OCKLEFORD

Structure and Function of the Crustacean Larval Salt Gland

FRANK P. CONTE

Pathways of Endocytosis in Thyroid Follicle Cells

VOLKER HERZOG

Transport of Proteins into Mitochondria

Shawn Doonan, Ersilia Marra, Salvatore Passarella, Cecilia Saccone,
                    and Ernesto Quagliariello

Contributors

Numbers in parentheses indicate the pages on which the authors' contributions begin.

FRANK P. CONTE (45), *Department of Zoology, Oregon State University, Corvallis, Oregon 97331*

SHAWN DOONAN (141), *Department of Biochemistry, University College, Cork, Ireland*

RICHARD E. FINE (1), *Departments of Biochemistry and Physiology, Boston University School of Medicine, Boston, Massachusetts 02118*

VOLKER HERZOG (107), *Department of Cell Biology, University of Munich, D-8000 Munich, Federal Republic of Germany*

ERSILIA MARRA (141), *Istituto di Chimica Biologica e Centro di Studio sui Mitocondri e Metabolismo Energetico del Consiglio, Nazionale delle Ricerche, Università di Bari, 70126 Bari, Italy*

COLIN D. OCKLEFORD (1), *Department of Anatomy, University of Leicester, Leicester LE1 7RH, England*

SALVATORE PASSARELLA (141), *Istituto di Chimica Biologica e Centro di Studio sui Mitocondri e Metabolismo Energetico del Consiglio, Nazionale delle Ricerche, Università di Bari, 70126 Bari, Italy*

ERNESTO QUAGLIARIELLO (141), *Istituto di Chimica Biologica e Centro di Studi sui Mitocondri e Metabolismo Energetico del Consiglio, Nazionale delle Recerche, Università di Bari, 70126 Bari, Italy*

CECILIA SACCONE (141), *Istituto di Chimica Biologica e Centro di Studi sui Mitocondri e Metabolismo Energetico del Consiglio, Nazionale delle Recerche, Università di Bari, 70126 Bari, Italy*

INTERNATIONAL

REVIEW OF CYTOLOGY

VOLUME 91

Membranes

INTERNATIONAL REVIEW OF CYTOLOGY, VOL. 91

Supramolecular Cytology of Coated Vesicles

RICHARD E. FINE* AND COLIN D. OCKLEFORD†

*Department of Biochemistry and Physiology, Boston University School of Medicine, Boston, Massachusetts, and †Department of Anatomy, University of Leicester, Leicester, England

I. Introduction

Historical—prior to 1975: Because of the small size of coated vesicles they were not described until the advent of thin sectioning techniques for electron microscopy. They were first observed in 1957 (1) in erythroblasts and during the next decade they were shown to be present in many different cells and tissues from animals and plants although several different names were used to describe them (2). It is likely that they constitute a ubiquitous and unique organelle in essentially all eukaryotic cells.

Very little biochemical investigation of coated vesicles was carried out prior to the seminal studies of Pearse in 1975 which we describe in the following section. Only two groups published partial purifications of coated vesicles before 1975. Kanaseki and Kadota (3) in 1969 isolated a partially purified preparation from guinea pig brain and demonstrated that the coat structure was a regular array of pentagons and hexagons producing an icosahedral structure. They also proposed that the closed lattice was formed from a hexagonal network by conversion of 12 hexagons to pentagons. The proposed lattice structure has been amply supported by Heuser's recent rapid-freeze deep-etch morphological studies (4); Scheide *et al.* (5) also partially purified CVs from chicken oocytes.

Even though almost all the early studies on coated vesicles were descriptive, the morphologists showed excellent intuition in proposing possible functions for coated vesicles. For example, in 1962 Palay (6) proposed that coated vesicles play an important role in membrane transfer processes. In 1964, Roth and Porter (7), in an extremely stimulating article, proposed that coated vesicles function in the uptake of specific extracellular proteins, based on their morphological observations of the uptake of yolk protein in forming mosquito oocytes. This study stimulated many articles on the role of coated vesicles in endocytosis in many cell types. Numerous articles suggested the possibility that coated vesicles are involved in secretory functions as well [see Kartenbeck (8)]. Finally, the work of Holtzman *et al.* (9) and Friend and Farquhar (10) indicated that coated vesicles are involved in the genesis of lysosomes.

II. Isolation of Coated Vesicles

Coated vesicles were first isolated to near homogeneity by Pearse in 1975 (11) using pig brain. The method took advantage of two properties characteristic of coated vesicles but not of other cellular organelles. Coated vesicles are quite small and relatively heterogeneous in size, ranging between 60 and 250 nm in diameter. Also coated vesicles have a high density compared to most other membrane organelles. It was claimed that about 75% of their mass is protein and 25% lipid (11), but this ratio must overestimate the protein content for the preparation tested contained a large number of empty cages without phospholipid.

Pearse's isolation is straightforward and relatively rapid, 24–30 hours from beginning to end. Briefly, it consists of a very vigorous homogenization of fresh brain (or many other tissues) in a Waring blender. The buffer used is 0.1 M morpholine-ethane sulfonic acid (MES) an effective buffer at 6.0–6.5, the pH region in which coated vesicles are most stable; 0.5 mM Mg^{2+} and 1 mM EGTA as well as 0.02% Na-azide are contained in the buffer.

After a 45-second homogenization, large membranes and other entities are removed by sedimentation at 20,000 g for 30 minutes. The relatively clear supernatant is then centrifuged at 100,000 g for an hour to pellet all remaining membranes. The resuspended pellets are then placed on a 5–60% sucrose gradient and centrifuged at 50,000 g for 2 hours. A white diffuse zone in the upper half of the gradient is removed. This area is enriched in coated vesicles. Many larger membranes such as microsomes and mitochondria are removed.

After pelleting and resuspension a 20–60% sucrose gradient sedimentation to near equilibrium is carried out. This lasts 16 hours, at 50,000 g. It results in the separation of the relatively dense coated vesicles (a bluish white opalescent band) and empty cages from the bulk of the smooth vesicles which appear white. This step seems to be the key purification step in the Pearse procedure. After another sedimentation the now relatively purified preparation is subjected to a third sucrose gradient centrifugation, onto 5–30% sucrose, for 1 hour at 100,000 g. The turbid white-blue zone in the top third of the gradient contains the most enriched coated vesicle population.

Pearse monitored the purification by examining negative stained samples using electron microscopy and was able to show the enrichment of vesicles surrounded by a regular lattice during the course of purification. When these purified preparations were subjected to SDS gel electrophoresis one major subunit polypeptide was present which was named "clathrin" from the Greek word meaning cage or bar.

Pearse's procedure and modifications thereof have been used to purify coated vesicles from many different tissues including liver (13), oocytes (14), adrenal medulla and cortex (16), placenta (17), skeletal muscle (18), and several types of cultured cells including lymphoma cells (15) and CHO cells (19). In all cases, the coated vesicles appeared similar in size and morphology. Also they all contain a similar 180,000 MW protein as the major component. The yield of coated vesicle protein obtained by this technique has been estimated to be about 3 mg/100 g (11,12,16).

During the last several years a number of alternative procedures for coated vesicle purification have also been developed. Perhaps the most widely employed is the procedure of Keen *et al.* (20). This technique employs the homogenization and centrifugation techniques of Pearse to obtain a crude membrane fraction. It then employs two sequential discontinuous sucrose gradients to isolate a partially enriched coated vesicle preparation. While this preparation is much cruder than that obtained by the Pearse procedure, the yield is much better and it serves as an excellent starting point in the purification of clathrin (21). Booth and Wilson (22) have used this method combined with treatment with wheat germ agglutinin to isolate a quite pure preparation of coated vesicles from placenta.

Recently, Pearse has proposed two methods which improve the isolation of purified coated vesicles from human placenta (23). One method consists of homogenization and centrifugation on isotonic buffer followed by a velocity gradient in 9–90% D_2O–H_2O. The coated vesicle enriched fraction from this gradient is then treated with 1% Triton which dissolves smooth vesicles while leaving the basket structure morphologically intact. Residual impurities are removed by a second D_2O–H_2O gradient. Alternatively the coated vesicles are purified by an equilibrium Ficoll–D_2O gradient. Pearse, in advocating this purification procedure, suggests that the very hyperosmolar environment to which coated vesicles are subjected when the conventional sucrose equilibrium gradient is used leads to both loss of vesicle contents and artifactual production of empty coats, devoid of membranes. Direct evidence of brain coated vesicle damage by high (40% or above) sucrose has recently been obtained by Nandi et al. (24). They substitute sedimentation through an 8% sucrose, 100% D_2O solution for the sucrose equilibrium gradient to obtain highly enriched brain coated vesicles. However, in comparing the various purification methods to isolate coated vesicles from rat liver and chick embryo skeletal muscle, we have found that the original Pearse procedure produces the best results both with regard to yield and purity (18,25). We see few or no empty cages in our preparations from these tissues. This suggests that brain coated vesicles are for some reason more susceptible to disruption by high sucrose than are coated vesicles from other tissues.

While treatment with 1% Triton X-100 as described by Pearse produces placenta coated vesicle lattices in which the contents remain somehow held in the reticulum left after removal of the phospholipids, no quantitative results were given for the amount of various content proteins lost after Triton X-100 extractions. Recently, quantitative estimates of insulin receptor and acetylcholinesterase loss have been carried out with coated vesicles purified from rat liver and chick muscle, respectively. It appears that extraction with 0.2% or higher levels of Triton almost quantitatively extracts these proteins while leaving the coat structure intact (R. Fine, unpublished). Similar results were obtained with G protein containing CHO cell coated vesicles (Rothman and Fine, unpublished results), It may be that for some as yet unknown reason Triton extraction removes some receptors and content proteins and not others, or alternatively coated vesicles from some tissues are more resistent to detergent treatment than others.

Recently, the use of isoosmotic Percoll gradients to replace the sucrose equilibrium gradient has also been reported (26).

Since coated vesicles purified by the sucrose gradient method are not completely homogeneous, containing some smooth vesicle contaminants, several procedures for further purification have been employed.

The use of glass bead permeation chromatography has been used as a final purification step for brain coated vesicles. This procedure separates coated vesicles from contaminating large membranes on the basis of size. Smooth vesicles of approximately the same size remain with CVs, however (27). Recently, chromatography on a S-1000 Sephacryl column which has a very high molecular weight exclusion limit has also been used for coated vesicle purification from yeast (28).

The fact that clathrin is negatively charged at pH 6.5 has been utilized to purify coated vesicles to near homogeneity by agarose gel electrophoresis (29). Since contaminating smooth vesicles, not containing clathrin, are less negatively charged, they migrate more slowly than do coated vesicles and hence can be separated completely. Coated vesicles from several different tissues including brain, liver, CHO cells (29), and skeletal muscle (18) have been purified to near homogeneity by this method. This technique up until now has been mainly useful as an analytic technique although recent experiments suggest that coated vesicles can be removed from the gel electrophoretically with a relatively high efficiency. The method may therefore be useful for preparative purposes (J. Squicciarini and R. Fine, unpublished). Recently, Merisko *et al.* (30) have reported the separation of brain coated vesicles from crude homogenates by immunoprecipitation with *Staphylococcus aureus* bound to anti-clathrin antibody. This technique may prove useful in isolation of coated vesicles from small amounts of material, e.g., cultured cells.

A. CHEMICAL AND IMMUNOLOGICAL STUDIES OF CLATHRIN

As Pearse first demonstrated (11,15) coated vesicles isolated from all animal tissues contain a major 180,000 MW polypeptide given the name clathrin. One-dimensional peptide mapping revealed that the peptide composition of clathrin isolated from human KB cells, pig brain, and bovine adrenal medulla were very similar (12). An antibody against bovine brain clathrin has been shown to specifically stain chick and reptile cells (31).

Very recently, coated vesicles have been partially purified from suspension cultured cells of tobacco (32). These coated vesicles, while appearing similar in fine structure and size to animal coated vesicles, contain a major polypeptide with a MW of 190,000 about 10,000 larger then animal clathrins. This finding suggests that clathrin while relatively conserved may have been significantly altered chemically when compared with the highly conserved actin and tubulin polypeptides. Interestingly, coated vesicles isolated from yeast, a lower plant, contain a major 180,000 polypeptide, identical in size to animal clathrin (28). Radioimmunoassays using anti-bovine brain clathrin also indicate that immunochemical differences are detectable between clathins from different tissues (33).

B. OTHER COATED VESICLE ASSOCIATED PROTEINS

Three major groups of polypeptides have been associated with purified coated vesicles isolated from brain and several other tissues. These groups have molecular weights of approximately 100,000, 50,000–55,000, and 35,000, respectively (29,14). The 35,000 doublet appears to be the clathrin light chains which will be discussed in the next section.

The 100,000 MW group of polypeptides seems to be associated with the vesicle since they sediment with the vesicles after 2 M urea treatment which solubilizes the coat (34). They are probably not intrinsic membrane proteins since they can be partially solubilized by extraction with 0.5 M Tris–HCl at neutral pH (20). Several functions have been proposed for this group of polypeptides. It was originally suggested that the 100,000 kd peptide(s) was associated with a coated vesicle calcium stimulated ATPase activity (34). However, it has been shown recently that essentially all the calcium ATPase activity can be separated from the coated vesicles by agarose gel electrophoresis and seems instead to be associated with presumed smooth vesicle contaminants (29).

The 100 kd polypeptides may play a role in the reformation of the clathrin coats after depolymerization (20). In this regard, there is recent evidence for a high-affinity clathrin receptor on uncoated vesicles, which is postulated to consist of the 100 and 50–55 kd polypeptides (35).

It has also recently been suggested that this group of polypeptides may be so-called ''adaptor'' molecules, serving as linkers between the clathrin lattice and the transmembrane proteins enclosed by the lattice (36).

There is also little evidence suggesting a role for the 50–55 kd polypeptides. Abstracts have been published claiming that tubulin (37) and glial filament subunits (38) may make up at least some of this group of polypeptides. Very recently, Pfeffer et $al.$ (39) have characterized a major 55,000 component of highly purified brain coated vesicles as α and β tubulin by several methods including peptide mapping and 1- and 2-D gel electrophoresis. Tubulin is also found associated with liver coated vesicles, but in much smaller amounts. The tubulin is dissociated from the coated vesicle by high salt treatment indicating it is not an integral membrane protein. As mentioned above the 55,000 peptide(s) has also been suggested to combine with the 100 kd protein to form the binding site for clathrin on uncoated vesicles (35).

Several other proteins are specifically associated with coated vesicles. For example, a protein kinase activity associated with brain coated vesicles was reported (26). This activity, which was not dependent on either Ca^{2+} or cyclic nucleotides, phosphorylates a 50,000 polypeptide which copurifies with coated vesicles on agarose gels.

Recently, a similar kinase has been identified in rat liver and chick muscle (40). Two 50,000 MW polypeptides have been shown to be phosphorylated on threonine residues and the two peptides have been identified as ATP binding sites by photoaffinity labeling with the ATP analog 8-azido-[α-^{32}P]ATP (41).

There is evidence that calmodulin, a ubiquitous calcium binding protein which activates many different enzymes including protein kinases, also copurifies with coated vesicles (42,42a). In this regard calmodulin may regulate the activity of a protein kinase which seems to be associated with the pair of 35 kd polypeptides shown to be the clathrin light chains discussed in the next section (43).

C. Coated Vesicle Structure

Clathrin Comprises the Protein Lattice

Pearse in her original report postulated that the 180,000 kd protein she named clathrin composed the major if not the sole component of the protein coat surrounding the vesicle (11). The experiments of Blitz *et al.* (34) demonstrated that the 180,000 kd protein could be selectively solubilized by treatment with 2 *M* urea. The other two major components at 100 and 55 kd remained with the uncoated vesicles. This finding indicated that the clathrin lattice was not intrinsically associated with the phospholipid bilayer. Other treatments such as 0.5 *M* Tris–HCl, increased pH, and lowered ionic strength also were shown to selectively solubilize clathrin (20,31,45). Pearse subjected the solubilized clathrin to chromatography and found that bands in the 35K region copurified with clathrin (46).

In 1978, several groups demonstrated that solubilized clathrin could reassemble into intact coats under relatively physiological conditions thus directly demonstrating that clathrin contained the necessary information to promote its own reassembly (21,47,20).

Various experiments suggest that polymerization occurs optimally at a pH between 6.0 and 6.5 in the presence of either Mg^{2+} or Ca^{2+} (45). The reaction occurs very rapidly and it appears to behave like a condensation process, where above a critical protein concentration (0.05 mg/ml) the clathrin self-associates to form the coat, in equilibrium with a small subunit pool.

A major breakthrough occurred in 1981 when two groups almost simultaneously demonstrated that the functional subunit comprising the clathrin lattice consists of a trimeric structure containing three 180,000 kd chains together with three light chains (48,49). The light chains have molecular weights of 33,000 and 36,000, respectively and are present in a 2:1 ratio. It has been demonstrated that one light chain is bound to each heavy chain (49).

D. The Triskelion

Ungewickel and Branton (48) made the important observation that the clathrin subunit when dried from a glycerol solution and viewed after low angle rotary platinum shadowing have a very distinctive structure (Fig. 1). These assemblies, which have a sedimentation constant of 8.4 S and a molecular weight of 630,000, appear to resemble a swastika with one leg missing and were given the

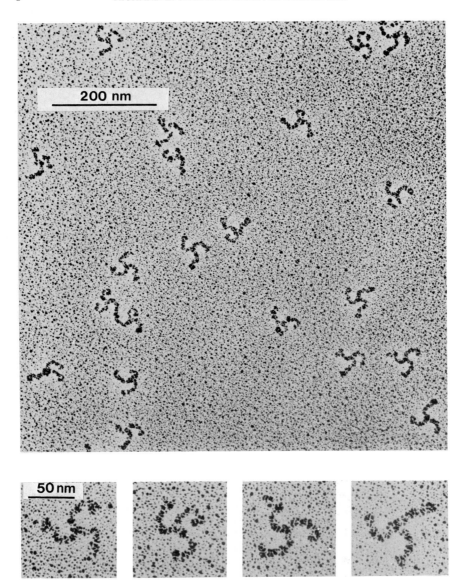

Fɪɢ. 1.   Low-angle-shadowing electron micrograph of purified clathrin assembly units. These 8.45 subunits are called triskelions after the symbol of 3 running legs used as the Isle of Man Crest. Micrograph from Ungewickell and Branton (48) with permission.

descriptive name "triskelion." Each clathrin monomer consists of one heavy and one light chain. The length of each leg is 45 nm with a clockwise bend approximately 20 nm from the central vertex.

The light chains appear to be essential for the proper *in vitro* reassembly of cages. Elastase treatment of triskelions removes the protease-sensitive light chains and also prevents normal cage assembly. It leads instead to large aggregate formation (49). Chymotrypsin produces the same result (50). Very recently it has been reported that whereas 60% of normal triskelions have all three legs bent in the same direction, elastase treatment reduced this to 30% (51). This finding suggests that the light chains may serve to orient the heavy chains properly so that correct polymerization can occur. Interestingly, trypsin treatment of the trimers, which removes the light chains and reduces the molecular weight of heavy chains to 110,000, by totally removing distal portion of the legs, does not block proper polymerization into cages as determined by negative staining (49).

The precise location of the light chains is not yet known with certainty. It has been reported that if the light chains are biotinylated, added back to the heavy chains and then probed with avidin conjugated to ferritin, the markers are localized near the vertex of the triskelion, suggesting the light chain is near this portion of the heavy chain (52). A direct visualization of several monoclonal anti-light chain antibodies confirmed this localization (53). This study also indicated that the two classes of light chains were distributed randomly in the triskelions.

The light chains can be readily purified to homogeneity by boiling crude coated vesicles preparation (43,52). Upon removal of the denatured protein by centrifugation only the light chains remain in solution as monomers. It appears that the extreme heat stability of the light chains is due to the high α helicity of the proteins indicating that they are long narrow molecules with a similarity to tropomyosin.

E. THE STRUCTURE OF THE COAT

In 1969 Kaneseki and Kadota (3) proposed that the coated vesicle coat was made of a protein lattice composed of a regular array of hexagons and pentagons. These studies were supported by work with isolated coated vesicles from brain and placenta (14,47,55,56). Crowther *et al.* proposed three structures for isolated brain coated vesicles (55) (Fig. 2). All three structures contain 12 pentagons as required for structures with icosahedral symmetry. Structures A and C contain eight hexagons and B four. Based on a molecular weight of 22×10^6 for the empty cages one might expect about 100–120 clathrin polypeptides of approximately 200 kd to comprise the cage. Since a complete cage in either structure A or C has 34 vertexes, this was consistent with three clathrins comprising each vertex or 102 clathrins in all, producing a molecular weight of 21×10^6. The discovery of the triskelion structure of the clathrin subunit supported that conclusion. From estimates of the size of an individual clathrin leg, 44.5 nm (48,57), and the measurement of the length of an edge of a coated cage, 19 nm (57), one can estimate that one leg contributes to two edges of the cage structure, with some 8 nm or 80 Å, left over.

Very recently, Crowther and Pearse (57) have proposed a detailed model of the cage structure based on negative stained preparations of triskelions, baskets,

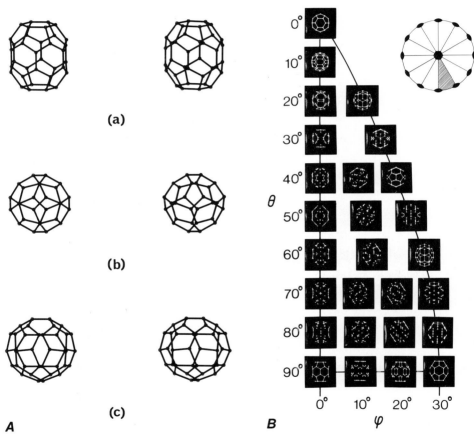

A   **B**

FIG. 2. (A) Stereopairs of 3 structures isolated from pig brain. Structure (a) consists of 8 hexagons and 12 pentagons; structure (b) comprises 4 hexagons and 12 pentagons; structure (c) contains 8 hexagons and 12 pentagons. (B) A gallery of projected images of structure (a) produced by computer graphical methods. The lattice is fitted and rotated on its axis by the number of degrees shown. The various projections may then be matched to the random images found in negatively stained preparations of isolated coated vesicles. From Crowther *et al.* (55) with permission.

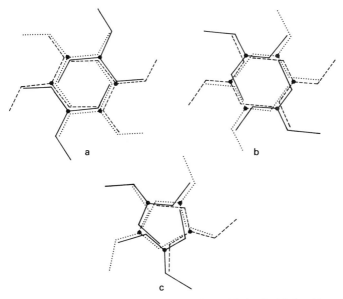

FIG. 3. Triskelion packing: (a) an hexamer of triskelions with simple side-by-side packing. (b) A hexagon formed by crossing over the legs; small rotations of the triskelions which were in essentially the same relationship as in (b) show how pentagons may form (c). After Crowther and Pearse (57) with permission.

and partially formed cages. The inner and outer segment of each leg contribute to two successive vertices being slightly rotated. Therefore each edge is formed from the proximal segments of two legs and the distal segments of two legs. The packing of a hexagon and pentagon illustrates how both can be formed similarly from triskelions with only a small distortion of the molecule (Fig. 3).

F. THE INTERACTION OF THE CLATHRIN LATTICE WITH THE VESICLE

There have been several studies suggesting an interaction between the clathrin lattice and the enclosed membrane. Unanue et al. (35) have recently demonstrated a high-affinity interaction between triskelions and uncoated vesicles. The interaction is protease sensitive. They suggest an interaction between clathrin and the core particles which are seen on the vesicle surface in negative stained EM. These particles are proposed to consist of the 100 and 55 kd peptides of the vesicles. Since, however, considerable clathrin remains associated with the uncoated vesicle, it is certainly possible that the interaction represents a clathrin–clathrin interaction.

Recently, it was reported that clathrin can directly interact with artificial phospholipid vesicles (58). Also it appears that clathrin can interact with the black lipid bilayer, inducing conductance changes in the bilayer.

Very recently, the same group has demonstrated that the interaction of clathrin with unilamellar dioleoyl phosphatidylcholine vesicles at pH 6.0 but not at pH 7.4 induces a very rapid vesicle fusion (59). This reaction occurs at a 1:500 ratio of clathrin to lipid. A number of other proteins (ovalbumin, IgG, trypsin, pronase, BSA, actin, tubulin, and synexin) did not produce any fusion at 10-fold higher concentrations.

Whatever the mechanism of clathrin:vesicle interaction, it is likely to involve the "foot" portion of the clathrin leg which does not seem to be part of the edges comprising the cage (57). This region, 8 nm in length, could certainly project onto the vesicle surface.

III. *In Vivo* Formation of Coated Vesicles

In their pioneering 1969 studies Kanaseki and Kadota (3) proposed that the coated vesicle is formed by a hexagonal lattice which is distorted so that pentagons are formed, inducing curvature in both lattice and underlying membrane. Ultimately the closed icosahedral lattice is produced, pinching off a piece of membrane and forming the icosahedral coated vesicle.

Ultrastructural studies on fibroblasts and other cells also suggest that the cytoplasmic fuzzy areas underlying the plasma membrane seem to have differing degrees of curvature consistent with various degrees of distortion in the hexagonal lattice (60).

Heuser's recent studies of coated vesicle formation in fibroblasts, employing replicas of extracted, deep-etched, and rotary shadowed cells indicate the basic correctness of this model (4). It is apparent from his beautiful micrographs (Fig. 4) that the first stage of CV formation is the assembly of a hexagonal lattice. The next stage of assembly appears to be the distortion of the lattices by the creation of a pentagon and heptagon from two hexagons. The heptagons apparently are then removed to an edge of the lattice by a series of distortions in the lattice. As more and more pentagons are produced the curvature of the lattice increases until the completed coated vesicle buds into the cytoplasm after 12 pentagons are formed. It would appear that energy assists some parts of coated vesicle formation *in vivo* since the rate of coated vesicle mediated endocytosis is greatly slowed when energy poisons are present or the temperature is reduced to very near 0°C (e.g., 36,60).

It has been found that one round of endocytosis of a ligand bound to the asialoglycoprotein receptor can occur in hepatocytes severely depleted of ATP (61). Receptor recycling is inhibited, however, indicating that this step is the one requiring ATP.

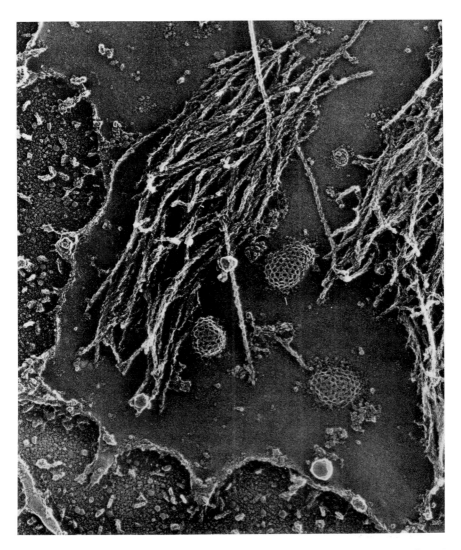

FIG. 4.   Quick-Freeze, deep-etch, rotary-replica electron micrograph of the internal surface of fibroblast plasma membrane associated with the substratum (in this case a coverslip). The polygonal clathrin lattice of 3 coated pits is clearly visible as are a number of sub-membrane F-actin filaments showing helical substructure. The area presented by this micrograph is about 1 by 2 μm. Micrograph from Heuser (4) with permission.

The mechanism(s) by which the distortions are produced in the hexagonal lattice are not known. Several groups have presented evidence consistent with the possible role of various cytoskeletal components in the formation of coated vesicles. As already mentioned actin, tubulin, and intermediate filament proteins have been reported to be associated with coated vesicles *in vitro* and *in vivo* (37,38,62). Fluorescent antibody studies indicate that coated vesicles and pits are not solubilized when cells are treated with detergents and appear therefore to be associated with the detergent resistant cytoskeleton (63).

Recently, several reports have appeared suggesting that Ca^{2+} is essential for coated vesicle formation *in vivo*. Maxfield *et al.* (64) found that coated vesicle mediated endocytosis of α_2-macroglobulin is blocked by chelation of extracellular Ca^{2+} with EGTA and suggest that Ca^{2+} is necessary for the clustering of transmembrane proteins in coated pits mediated by crosslink formation catalyzed by transglutaminase, a Ca^{2+}-requiring enzyme (64,65).

There is also evidence that calcium interacts with calmodulin bound to the coated membrane and in some way aids the formation of coated vesicles. Salisbury *et al.* have published evidence that trifluoperazines, relatively specific inhibitors of calmodulin-associated processes, block receptor mediated endocytosis of IgM by cultured lymphocytes (59,62,66).

As discussed in the previous section, very recently Lisanti *et al.* have presented evidence that the clathrin light chains bind calmodulin in the presence of Ca^{2+} (43). They also present evidence that the light chains have a kinase activity causing phosphorylation of several coated vesicle polypeptides including clathrin (44). Antibodies directed against the light chains block this phosphorylation. They speculate that phosphorylation is involved in coated vesicle assembly and/or disassembly.

A. Do Coated Vesicles Really Exist in the Cell?

Recently Pastan and associates have proposed that coated vesicles do not exist, and that just before the vesicle buds from the membrane into the cytoplasma, the clathrin lattice somehow slips away from the vesicle, remaining on the membrane (67). They present two types of evidence to support this claim. One is that many apparent coated vesicle profiles in the cytoplasm near the plasma membranes of mouse 3T3 fibroblasts can actually be shown to be connected with the plasma membrane by a slender stalk when serial sections are examined (68). Also they microinject an anti-clathrin antibody into the cell, and it appears that even though the antibody binds to coated pit regions in the cell, receptor-mediated endocytosis is not prevented (68). They postulate that isolated coated vesicles are actually artifactual, representing coated pits that form during the isolation procedure.

Besides the obvious conceptual difficulty in proposing a mechanism by which a nearly enclosed clathrin lattice could somehow slide away from the closing vesicle surface, there is now direct evidence supporting the existence of coated vesicles. Very recently Fan *et al.* (69) have shown that successive 60 nm serial sections through 3T3 L1 cells incubated with cationic ferritin, a widely utilized marker for absorptive nonspecific endocytosis, for 2 minutes at 37°C, revealed that 47% of the apparent coated vesicles treated just beneath the cell surface were true coated vesicles, whereas the other 53% were coated pits. A very recent study of L cells and human fibroblasts also showed that true coated vesicles are intermediates in the absorptive endocytosis of cationic ferritin (70). True coated vesicles have also been shown to exist in CHO and skeletal muscle cells utilizing serial sections (D. Sack, R. Benson, and R. Fine, unpublished results).

B. Uncoating a Coated Vesicle

Even though it is reasonably certain that coated vesicles do in fact exist, most investigators agree that the coat of the coated vesicle is removed very rapidly, probably within a minute after coated vesicle formation (e.g., 36,60). Since the coated basket would be expected to be a favored structure from a thermodynamic standpoint as indicated by the favorable equilibrium for cage formation *in vitro* under relatively physiological conditions, it is likely to require energy to remove the coat. Recently, Patzer *et al.* have demonstrated that a partially purified extract of liver cytoplasm catalyzes the removal of the coat from coated vesicles in the presence of ATP (71). Trimers apparently are the product of this reaction.

The uncoating enzyme has been purified to homogeneity (72). It is a dimer composed of two identical 70,000 MW polypeptides. It has an unusually high affinity for ATP, $K_d \sim 0.1$ μM. It appears to form a stoichiometric 1:1 complex with the clathrin trimers, i.e., 3 mol of uncoating dimers per 3 mol of clathrin. This complex is incapable of reforming clathrin cages.

IV. Functions for Coated Vesicles

A. Receptor-Mediated Endocytosis

Roth and Porter (7) in 1964 first suggested the role of coated vesicles in receptor-mediated endocytosis of ligands into cells. Although some work was carried out during the next 10 years suggesting the generality of their findings, it was the now classic experiments of Anderson *et al.* which thrust receptor-mediated endocytosis into the forefront of cell biology (60). Using LDL labeled with ferritin they demonstrated, using transmission EM, that one can describe receptor mediated endocytosis as a series of discrete steps.

Step 1 is the specific interaction of the ligand with a receptor, a protein anchored in the plasma membrane. In the case of the LDL-receptors 78% of them are clustered over specific dimples in the plasma membrane with a cytoplasmic clathrin coat. Several other receptors have been shown to be preclustered over coated pits using transmission EM including transferrin receptors in HeLa cells (73), and asialoglycoprotein receptor on hepatocytes (74,75). On the other hand some receptors are not clustered to a significant extent in coated pits until bound to ligand molecules (69,76,77). Direct evidence that several different receptor–ligand complexes can be included in a single coated pit has been obtained (76,78). There is now evidence that the receptor ligand interaction can actually trigger coated pit formation. Salisbury et al. (62) have found evidence for transmission electron microscopy (TEM) that IgG binding to receptors on human lymphoid cells leads to a rapid 3-fold increase in coated pits. Connolly et al. (79,80) have found a similar 3-fold increase in coated pits following NGF or EGF binding to their respective surface receptors on PC12 cells, a human pheochromocytoma derived cell line, using both TEM and scanning (EM). Anderson et al. (81) have used cells from a patient with familial hypercholesterolemia to demonstrate the importance of receptor clustering over coated pits for efficient ligand endocytosis to occur. In fibroblasts from this patient, the cells bind normal quantities of LDL; however, because the receptors are not associated with coated pits either before or after ligand binding, the LDL is internalized and degraded at a much slower rate than normal, leading to aberrant cholesterol metabolism and early atherosclerosis. Presumably there is a special conformation associated with receptors and other proteins which cluster in coated pits which allow them to be recognized by the coat structure. Presumably many plasma membrane proteins which are destined to remain permanently embedded in the PM do not have this conformation. Two plasma membrane proteins, ϕ and H63 antigen, have recently been demonstrated to be excluded from coated regions (82), as has receptor bound thrombin (82a), and it is likely that many more proteins will be found to be in this category. In this regard, it has recently been found that luteinizing hormone receptors on ovarian granulosa cells are internalized at least six times slower than bulk membrane, which can be quantitatively accounted for by coated vesicle-mediated endocytosis (31,83). This receptor, therefore, apparently is excluded from coated pits. There is evidence that cholesterol is also at least partially excluded from coated pits (84).

Step 2 is the formation of coated vesicles, removing ligand and receptor from the plasma membrane. Very soon after binding, within 30 seconds, Anderson et al. were able to visualize the ferritin LDL in coated vesicles presumably totally separate from the plasma membrane (60).

Recently coated vesicles were isolated from adrenal cortical cells, which have a high number of LDL receptors in a cryptic form, i.e., only available for LDL binding after detergent treatment (16). Whether or not these coated vesicles were engaged in endocytosis was not determined.

Step 3: The smooth vesicle intermediate. Very rapidly after the coated vesicle buds from the plasma membrane, probably within a minute, the coated vesicle loses its coating, either partially or fully. At this point the ligand seems to lose its affinity for its receptor as detected in thin section because there is no longer close apposition between the ligand and the membrane (74,85).

Very recently Geuze *et al.* (85) have demonstrated that an organelle they call CURL, an acronym for compartment of uncoupling of receptor and ligand and which almost certainly corresponds to the smooth vesicle intermediate described by others, is the organelle in which the asialoglycoprotein receptor and its ligand separate. As Fig. 5 demonstrates, the ligand appears in the lumen of CURL, apparently no longer in contact with the membrane. The receptors are now found inside the tubular extensions of CURL. Interestingly, clathrin is also concentrated in the cytoplasm surrounding the tubular extension, suggesting that coated vesicles may also be an intermediate in receptor recycling to the plasma membrane.

Anderson *et al.* (86) have used ferritin-labeled anti-LDL receptor to directly demonstrate that the receptor enters the cytoplasm with the ligand and then separates from it and recycles back to the plasma membrane by an as yet unknown route. There is also evidence for recycling of transferrin receptors (73), asialoglycoprotein receptors (87), and insulin receptors (88). Presumably a compartment very similar to CURL is the site of receptor–ligand separation in these systems as well.

Step 4: The fusion of ligand containing smooth vesicle with lysosomes. In the case of LDL, Anderson *et al.* (60) were able to visualize LDL in multivesicular bodies, a particularly distinctive class of lysosomes, within 10 minutes after binding and warm up.

B. VARIATION ON A THEME: DIFFERENT PATHWAYS FOR ENDOCYTOSIS

Is receptor-mediated endocytosis unique or do other "nonspecific" ligands actually go through the some pathway, i.e., coated pit → vesicle → endosome → lysosome.

Pastan and colleagues have proposed a unique smooth vesicle, given the name "receptosome," which contains the receptors for various hormones and other receptor specific ligands, as a relatively long lived intermediate in receptor-mediated endocytosis (67). It presumably corresponds to the smooth or uncoated vesicle intermediate first described by Anderson *et al.* (60). The evidence supporting the "receptosome" has recently been reviewed (67).

A major postulate of the "receptosome" hypothesis, which states that the receptosome contains only ligands and their receptors and not other "nonspecific" substances, was contraindicated by results of Ryser *et al.* (89). Using horse radish peroxidase (HRP), a well-established marker for fluid phase endo-

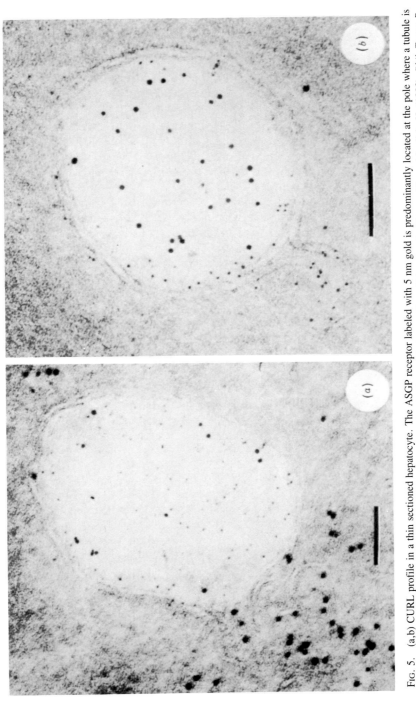

Fig. 5. (a,b) CURL profile in a thin sectioned hepatocyte. The ASGP receptor labeled with 5 nm gold is predominantly located at the pole where a tubule is connected. Most of the ligand, labeled with 8 nm gold is present free in the vesicle lumen. From Schwartz, A., Geuze, H., and Lodish, H. (1983). *Phil. Trans. R. Soc. London Ser. B* **300**, 229–235, with permission.

cytosis, and poly L-lysine conjugated HRP, a substance which binds tightly to the plasma membrane and is considered a membrane-marker for absorptive, nonreceptor-mediated endocytosis, they have shown in cultured fibroblasts that both substances are internalized and degraded mainly via the coated vesicle → smooth vesicle → lysosome route defined by Anderson *et al.* (60). Both substances are seen in intracellular structures described as "receptosomes" by Pastan and Willingham (67). It has also recently been demonstrated that when 3T3L1 cells are incubated for a short time, 2–5 minutes, with cationic ferritin which is another widely used marker for absorptive, nonspecific endocytosis, a large smooth vesicle is maximally labeled (69,70). It is therefore likely that any form of endocytosis in fibroblasts and other "nonspecialized" cells will mainly involve the above pathway. As we will shortly see, however, cells with more specialized functions, e.g., secretory and epithelial cells, may have much more complex pathways.

The receptor-mediated pathway for endocytosis appears also to be followed in fibroblasts as well as in tumor cells by epidermal growth factor (EGF), a widely studied polypeptide hormone. Morphological studies (90) using EGF-ferritin have demonstrated that EGF is internalized and degraded in the same manner as LDL. This view has been confirmed by biochemical studies in which ^{125}I-labeled EGF has been localized in intracellular structures with the characteristics of coated vesicles and lysosomes following receptor binding at 4°C and warm up (91). One major difference between the LDL and EGF pathway relates to the fate of the internalized receptor. While the LDL receptor rapidly separates from the ligand after internalization and is recycled to the plasma membrane (86), the EGF receptor appears to be degraded in the lysosome concomitantly with the bound ligand (92). This process, known as receptor down-regulation, is known to occur with other hormone receptor systems as well, most notably with the insulin:insulin–receptor system (93).

Besides hormones, two other groups of substances, viruses and polypeptide toxins, seem also to invade the cell through the same pathway. Perhaps the best studied case is that of the enveloped RNA virus, Semliki Forest Virus (SFV). Morphological evidence indicates that this virus traverses the coated vesicle pathway, before fusing with the lysosome or some other acidic vesicle thereby releasing the infectious nucleocapsid into the cytoplasm (94). The evidence suggesting that many other viruses as well as diphtheria toxin and other bacterial toxins and toxic plant lectins such as ricin follow the same pathway is more indirect, but similarities in kinetics of internalization and also the effects of lysosomotropic agents on these substances suggest the involvement of the coated vesicle pathway (95–97). There is, however, contradictory evidence as well (98,99).

The most striking effect of inhibitors on various endocytic processes is seen with lysosomotropic reagents such as NH_4Cl and other basic amines and chloroquine. These agents are all known to raise the pH of the lysosomes thereby inhibiting lysosomal enzymes which require acidic pH (100). There is also evidence that these reagents may block lysosomal fusion with other vesicles, possibly because an acid pH is necessary for fusion to occur. Na/H$^+$ ionophores such as monensin seem to have similar effects (101,102). These compounds interfere with the degradation of LDL and many peptide hormones (e.g., 91,101,103) and also block the infectivity of several viruses and toxins (e.g., 90,97,102,104).

It has been claimed that amines as well as certain other agents specifically block the internalization of receptor bound ligands, possibly by interfering with the action of transglutaminase which was postulated to covalently cross-link receptors in coated pits (105). In view of the fact that almost all the reagents postulated to block internalization directly are basic compounds which will affect the pH of lysosomes and other acidic compartments, it is likely the effects on internalization are secondary to the primary inhibition of acidification. It has been demonstrated that coated vesicles accumulate in cultured muscle cells treated with chloroquine (106). NH$_4$Cl has been shown to block EGF degradation and to lead concomitantly to the accumulation of EGF in intracellular structures which behave biochemically like coated vesicles as well as on plasma membranes (91). A reasonable interpretation of these experiments is that the coated vesicle-mediated endocytosis pathway is tightly coupled and a blockage at the final stage, i.e., lysosome functions leads to the blockage of the whole system. Other support for the idea of a tightly coupled pathway comes from recent data on LDL uptake. Monensin, a Na$^+$H$^+$ ionophore known to disrupt the Golgi apparatus and block secretory pathways also blocks LDL receptor recycling trapping receptors and ligand in an internal vesicular compartment (101). Interestingly 50% of the receptors still remains at the cell surface suggesting again that the endocytosis system is a tightly coupled one (86).

Most investigators have assumed until recently that the lysosome is the only acidic compartment in the endocytosis pathway, and hence the site of amine inhibition and therefore the compartment from which viruses, toxins, and possibly hormones as well escape to the cytoplasm. Recent evidence obtained by Tyco and Maxfield (107) indicates that the prelysosomal endocytotic vesicle corresponding to the endosome or CURL is also acidic, thereby allowing the above mentioned substances to escape to the cytoplasm before encountering the hazards of the lysosome. Very recently, direct evidence that an acidic compartment is the compartment from which SFV escapes to the cytoplasm was obtained (108). It was first shown by Dunn et al. (109) that incubation at 20°C does not inhibit endocytosis but blocks endosome, lysosome fusion. Making use of this finding, Helenius and associates (108) demonstrated that SFV could infect cells at 20°C, thereby suggesting that the acidic prelysosomal compartment, probably corresponding to CURL, is indeed the vesicular compartment from which SFV escapes. Consistent with this is their observation that at 20°C the endocytosed SFV accumulates in a light density membrane which has no lysosomal enzymes. Other ligands and VSV virus also have recently been shown to accumulate in this compartment (110,111).

Galloway *et al.* (112) have very recently isolated endosomes from cultured macrophages and fibroblasts which contain internalized fluorescein-dextran, a pH-sensitive fluorochrome. They have demonstrated that the addition of ATP leads to the acidification of the endosomes by approximately 0.5–0.8 pH units, providing more evidence consistent with the existence of acidic endosomes in cells.

Very recently, two groups have independently demonstrated that highly purified preparations of brain coated vesicles contain an ATP-dependent proton pump, which causes the isolated coated vesicles to become acidified with respect to the surrounding solution (113,114). The pump activity is not inhibited by agents which interfere with the well characterized Na^+–K^+ and Ca^{2+} ATPases as well as the mitochondrial ATPase. It seems to bear similarity to the recently described lysosomal ATPase (115). It is likely therefore that this pump is responsible for the acidification which occurs rapidly after endocytosis. Consistent with this is the recently obtained evidence that the endosome and lysosome contain very similar H^+ ATPases (116).

Recently, Larkin *et al.* (117) have demonstrated that a depletion of intracellular K^+ causes a reversible decrease in coated pit formation and inhibits receptor-mediated endocytosis and fluid phase endocytosis as well. Interestingly, LDL receptors remain in clusters during this treatment suggesting that these receptors remain clustered even in the absence of coated pits.

C. ENDOCYTOSIS IN SPECIALIZED CELLS

A tremendous volume of data has accumulated in recent years on the role of coated vesicles in endocytic processes in various specialized vertebrate cells as well as in lower organisms. It is beyond the scope of this review (or these reviewers) to describe all of this research. We have therefore selected certain cell types and specific processes for the following discussion, mainly with two goals in mind. (a) The illustration of variations from the generally described pathway discussed above. (b) To indicate systems where both morphological and biochemical methods has been employed, at least theoretically opening up areas for a molecular approach.

Endocytosis in Liver via Coated Vesicles

Even though liver is most commonly regarded as a secretory organ, an interesting array of endocytotic processes also occur. We will concentrate on two involving hepatocytes. (a) Asialoglycoprotein uptake and degradation and (b) insulin internalization and degradation.

Hepatocytes contain large numbers of asialoglycoprotein receptors on their sinusoidal surface. These receptors recognize galactose residues on asialoglycoproteins, and the bound proteins are efficiently internalized and degraded. The receptors seem to be recycled to the plasma membrane after internalization similarly to the LDL receptor (75,87). It has been found that some asialoglycoprotein receptors are clustered in coated pits (74,85).

Recently, a smooth vesicle fraction has been isolated from rat livers perfused for 10 minutes with ^{125}I-labeled asialotransferrin (118). These vesicles surprisingly appear to have the asialoglycoprotein receptors facing the cytoplasm suggesting that the receptor orientation becomes inverted at some point following internalization. Blumenthal et al. (119) have data indicating that the orientation of the asialoglycoprotein receptors can be inverted in a model membrane by changing the electrical potential of the membranes and they speculate that vesicle acidification after internalization could trigger such an inversion.

Hubbard and associates have recently studied the fate of ^{125}I-labeled EGF and ^{125}I-labeled ASOR, a ligand internalized via the asialoglycoprotein receptor (ASR) and their respective receptors (120). While both ligands are degraded after internalization most of the EGF-Rs are also degraded while all the ASRs are recycled. Incubation of liver at 16°C results in the accumulations of both ligands in a light membrane compartment, presumably CURL. Interestingly the ASRs are rapidly returned to the cell surface while the EGF-Rs are not. These results also indicate the existence of multiple pathways for different receptors and ligands in the same cell.

Insulin receptors are also found on the hepatocyte sinusoidal surface and are thought to be internalized following insulin binding (69). Recently, Fehlman et al. (88) have demonstrated by photocoupling an insulin derivative to insulin receptors that some insulin receptors appear to be recycled to the plasma membranes, Pilch et al. (121) have recently demonstrated that purified rat liver coated vesicles contain latent insulin receptors. Crosslinking studies reveal that the insulin receptors have the same $\alpha_2 \beta_2$ subunit structure as do plasma membrane receptors. By isolating coated vesicles from rat livers perfused for 3 minutes with either ^{125}I-labeled insulin alone or in combination with excess cold insulin it was demonstrated that there is a very rapid specific receptor-mediated association of insulin with coated vesicles. Recent EM autoradiographic studies using 3T3L1 cells incubated for increasing time with ^{125}I-labeled insulin also indicated an initial preferential localization of the ligand to microvilli, followed by its rapid translocation to coated pits and coated vesicles (69). Both studies indicated a very rapid association of insulin with coated vesicles, within 2–5 minutes after incubation, followed by an equally rapid transfer to another, presumably smooth vesicle intermediate. These data are consistent with a major role for coated vesicles in the internalization of insulin in hepatocytes.

D. TRANSEPITHELIAL TRANSPORT VIA COATED VESICLES

Another coated vesicle-mediated endocytic pathway has been studied by numerous workers in several types of specialized epithelial and endothelial cells. The human placenta has been used as a vaulable model. Among its numerous functions the placental trophoblast cells transfer several types of proteins specifically from the maternal to the fetal circulation including IgG, transferrin, and ferritin. Several groups have demonstrated morphologically that there are many coated pits and coated vesicles present in placenta (122–124). Using ferritin-labeled IgGs it has been suggested that coated vesicles containing specific proteins bud from the maternal surface. Whether these traverse the cell intact, without losing their coats, and fuse with the fetal surface plasma membrane, thereby releasing their content of proteins into the fetal circulation, is not yet clear. Moreover three groups have recently isolated coated vesicles from placental cells which contain IgGs, transferrin, and ferritin (22,23,125).

Another tissue which has been studied in some detail is the neonatal intestinal epithelium. Tracer studies used HRP or ferritin-labeled IgGs (126) reveal that tubular elements lying at the base of the microvilli lining the apical surface of the intestine endocytose the IgG and HRP. The IgG is transferred to coated vesicles while the HRP is transferred to lysosomes and degraded. The coated vesicles traverse the cell intact, then fuse with the basolateral plasma membrane of the cells releasing the IgG molecules into the fetal circulation. The IgG receptors appear well adapted for this task, having high affinity for the IgG molecule at the rather acidic pH of the intestine and a relatively low affinity at the neutral pH of the blood on the basolateral surface (127).

It would appear from these results that the cell can distinguish between receptor bound ligands and other contents and can separate the two by as yet unknown means. Presumably this process may be analogous to the more classical endocytosis pathway in that a separation of receptors and contents occurs; however, in the transcellular pathway receptor bound ligands remain on the receptor while in the classical pathway they separate. Very recent studies of transport of colloidal gold-labeled IgG across the rat yolk sac *in vitro* also suggest coated vesicle mediated transcellular protein transport (128).

Another interesting endocytosis system is the transferrin–transferrin receptor system which is crucial for iron transport into cells. Transferrin, a serum protein with a very high affinity for Fe^{3+} at neutral pH, binds to its receptor, a 180,000 S-S linked dimer of two identical 90,000 subunits (129,130). The receptors are clustered in coated pits (73). The receptor–ligand complexes are internalized via coated vesicles, reach an acidic compartment where the iron, now having essentially no affinity for transferrin, is released (131). The transferrin and the receptor are then recycled back to the cell surface and the transferrin released because the apoprotein no longer has an appreciable affinity for the receptor (132,133).

Recently, the intriguing finding has been made that asialotransferrin seems to be partially resialylated during its traverse through the cell (134). Since the Golgi apparatus is thought to be the major if not the sole site of sialylation, one can suggest that the Golgi apparatus may lie on the membrane receptor recycling pathway as has been suggested by several morphological studies (135,136).

Bliel and Bretscher (73) have recently done quantitative studies on the amount of external and internal transferrin receptors in HeLa cells. In contrast to the LDL receptors which are mostly found at the cell surface (86), only 25% of the transferrin receptors are at the surface at any one time. It appears also to take 20 minutes for an individual transferrin receptor to recycle back to the cell membrane, while both LDL and asialoglycoprotein receptors require only 4 minutes (75,86). Among other possibilities these results may suggest that there are multiple membrane recycling pathways in cells.

E. Phagocytosis

Very recently Aggeler and Werb (137) have visualized the initial events during phagocytosis of latex beads by cultured macrophages using both conventional thin-section transmission EM and the high-resolution platinum replica EM technique developed by Heuser (4). These studies demonstrate the association of large areas of clathrin basket work on the cytoplasmic face of nascent phagosomes. Very recently these workers have also incubated macrophages on a flat surface containing a high concentration of IgG (138). The macrophages spread completely on this surface; presumably this spreading is mediated by the binding of IgG by the cells' Fc receptors. The cytoplasmic surface of the cell overlying the substratum become densely coated with clathrin basketworks. These studies taken together suggest an important role for clathrin in phagocytosis as well as in cell attachment and spreading on surfaces. Evidence that clathrin basketworks are also involved in cell–substrate interactions in Hela cells has very recently been reported (139).

F. Endocytosis and Membrane Recycling

The concept of specific coated vesicle-mediated membrane recycling was developed mainly by Heuser and Reese (140) to explain the ability of the presynaptic endings of frog neuromuscular junctions to very rapidly replenish their stock of synaptic vesicles following a short burst of electrical activity. At very early times (milliseconds) after the stimulus is given all the synaptic vesicles fuse with the plasma membrane causing a large (several fold) increase in the surface membrane. Very rapidly thereafter a large increase in the number of coated vesicles is seen followed sequentially by the formation of large cisternal elements and then by the appearance of synaptic vesicles. Because of the fact that synaptic vesicles contain larger intramembranous particles (IMPs) than does the presynaptic plasma membrane, these particles could be directly followed by freeze fracturing (141). It was demonstrated in this way that following the addition of large IMPs to specific "active zones" during exocytosis and transmitter release, the particles apparently diffuse away from these zones, forming microaggregates which then are selectively removed by coated vesicles budding from the plasma membrane into the cytoplasm (142). It appears also that bulk lipid-rich particle-poor areas are rapidly removed from the plasma membrane by large noncoated vesicles.

Because of the selectivity of the coated pit illustrated by receptor-mediated endocytosis, it becomes reasonable to view this membrane recycling as yet another example of a selective coated vesicle-mediated membrane transfer. It is very likely that other selective membrane retrieval processes which occur in many secretory cells, e.g., pancreas and parotid gland, are also mediated by coated vesicles. Studies showing increased numbers of coated vesicles near the plasma membrane of pancreatic and adrenal medulla cells stimulated to secrete are consistent with this possibility (143,144), as is a study of postfertilization events in the sea urchin egg (144a).

Recently, Bretscher (145,146) has made use of giant HeLa cells to demonstrate that three different receptors, for transferrin, LDL, and ferritin, respectively, are internalized, after ligand binding, via randomly distributed plasma membrane coated pits, but are recycled to the cell surface at the leading edge of the cell. This result indicates the existence of a discrete directional component for exocytotic membrane flow.

Bretscher has proposed that coated vesicles may also serve to recycle phospholipid molecules from one membrane to another (147). This is almost certainly the case in the fibroblast which completely turns over its membrane constituents every 60–100 minutes (148) and in which the coated vesicles represent the major vesicle for endocytosis (60,89). There may be some selectivity with regard to coated vesicle-mediated lipid transport as well. There is some morphological evidence that coated pits and vesicles may be low in cholesterol (80), although highly purified brain coated vesicles have been shown to contain a sizable amount of cholesterol (11). If coated pits were able at least partially to exclude cholesterol it could help to explain how the plasma membrane cholesterol:phospholipid ratio is maintained at a much higher value than that of internal membranous organelles.

V. Formation of Lysosomes and Recycling of Lysosomal Enzymes

As early as 1963, Palay suggested that coated vesicles may play a role in the genesis of lysosomes (149). Numerous cytochemical studies employing specific staining for acid phosphatase, a known lysosomal marker enzyme, have shown that many coated vesicles and coated pits mainly localized around the Golgi apparatus and Gerl contain this enzyme (9,10). More recent studies have localized other lysosomal enzymes in coated vesicles (150). Coated vesicles are therefore considered to be a class of primary lysosomes.

Numerous studies carried out during the last 10 years have shed considerable light on the biogenesis and transport of lysosomal enzymes. Lysosomal enzymes are synthesized in the RER, and transported to the Golgi apparatus where their high mannose type oligosaccharides are modified by a two step process to contain one or more mannose 6-phosphate groups (151). These groups presumably allow the enzymes to be recognized as destined to be transferred to lysosomes since lack of this marker in patients with I cell disease results in the secretion of these enzymes instead of transport to the lysosomes (152).

Recently, the mannose 6-phosphate receptor was shown to be present on the cell surface as well as in internal organelles including ER and Golgi (153,154). Very recently Willingham et al. (155) using light and EM cytochemical techniques have demonstrated that exogenously added lysosomal enzymes carrying mannose 6-phosphate bind to the receptor on the cell surface, cluster in coated pits, and get internalized and transferred to lysosomes by the normal receptor mediated pathway.

The mannose 6-phosphate receptor has been purified from liver and other tissues and cells (156,157). It is a 215,000 MW glycoprotein with an acidic p*l*. This receptor has recently been demonstrated to be present in highly purified coated vesicles isolated from rat liver and bovine brain (158). It is latent, i.e., binding to ^{125}I-labeled lysosomal enzymes is stimulated by addition of detergent. It appears also that various lysosomal enzymes are present in a latent form in the liver coated vesicles which can be displaced in the presence of detergent by the addition of mannose 6-phosphate. Interestingly, not all the enzymes are displaced suggesting that other receptors may also be involved.

Very recently, Campbell and Rome (159) have demonstrated the presence in agarose gel purified liver and brain coated vesicles of six different lysosomal enzymes. Treatment of detergent disrupted liver coated vesicles with mannose 6-phosphate specifically displaced 80% of the cryptic enzyme activities. When highly purified liver coated vesicles and lysosomes were treated with anti-β-hexosamididase and anti-β-galactosidase and resultant immunoprecipitates analyzed on SDS gels, high-molecular-weight bands were present in the immunoprecipitates from coated vesicles which were not present in lysosome immunoprecipitates. These data suggest that coated vesicles are transporting newly synthesized lysosomal enzymes to the lysosomes. Since the addition of the phosphate group occurs in the Golgi apparatus it is reasonable to conclude that coated vesicles containing the lysosomal enzyme precursors are of Golgi or post-Golgi origin.

Although these studies have confirmed the early theories concerning the role of coated vesicles in lysosome biogenesis, we still do not know precisely at what stage(s) of the process they are involved. In fact, the precise route of lysosomal enzyme transport is controversial with respect to whether the enzymes are transported directly from Golgi to lysosome (154) or whether they first are transported to the plasma membrane and then back to the lysosome (152).

VI.   The Roles of Coated Vesicles in the Treatment of Newly Synthesized Secretory and Plasma Membrane Proteins

Studies carried out over the last 30 years by Palade and colleagues (reviewed in 160) indicate that proteins and glycoproteins destined for secretion are synthesized, transported, and sequestered in membrane-enclosed compartments. These compartments were elucidated in a brilliant series of morphological and biochemical experiments and shown to consist of separate membranous elements connected in series. The newly synthesized proteins are first inserted into the rough ER during synthesis, transported to the Golgi apparatus in small transitional vesicles, and from the Golgi into secretory vesicles which ultimately discharge their contents by an exocytotic fusion with the plasma membrane. The secretory vesicle's membrane becomes part of the bulk plasma membrane during exocytosis.

Palade in his Nobel prize address (160) proposed that proteins destined for deposition in the plasma membrane may follow the same pathway as do secretory proteins. This has now been shown to be the case with several plasma membrane proteins. Perhaps the best studied is the surface glycoprotein (G protein) of the vesicular stomatitis virus. This protein has been shown by both biochemical and morphological techniques to travel sequentially from ER through transitional vesicles to Golgi and hence through another population of vesicles to the plasma membrane (161,162). A very considerable selection problem is involved in determining which particular proteins are to be removed from the ER. Rothman (163) has recently estimated that only 0.01% of the protein in the ER at any time is destined for removal. It is reasonable, therefore, to postulate a role for coated vesicles in this process since they have been demonstrated to play the role of a selective filter in endocytotic process.

  Much morphological evidence has accumulated indicating the role of coated vesicles in the transport of a variety of secretory proteins in various animal and plant cells (reviewed in 8). These studies have relied on the recognition of specific secretory products due to their characteristic appearance. Some of these include pancreatic secretions (164), prolactin secretion in rat pituitary (2), casein aggregates in secretory granules in mammary epithelium (165), lipoprotein secretion by hepatocytes (166), and cell wall material containing secretory granules in plants (8). These studies when taken together suggest a role for coated vesicles in the terminal phase of secretion, that is between the Golgi apparatus and plasma membrane. Palade and Fletcher (167) have also suggested a role for CVs as the transitional vesicles transferring newly synthesized proteins between ER and Golgi. A problem with all these studies is that it could not be ascertained with certainty whether or not the presumed secretory marker was in fact en route to the plasma membrane or represented endocytosed material.

A recent immunocytochemical study has been carried out which overcomes the above mentioned objections. Developing chick myotubes in culture have been shown to synthesize and transport acetylcholine receptors (AChR) to the plasma membrane according to the classic ER → Golgi → PM route (168). Bursztajn and Fischbach (169) have used HRP labeled α-bungarotoxin (αBGT) a specific AChR antogonist which is very tightly, almost irreversibly bound to AChRs to localize AChRs in cells. They demonstrate that coated vesicles in saponin-permeabilized muscle cells contain AChR as shown by HRP histochemistry after incubation with HRP-αBGT. The function of these was determined by preincubating the cells with αBGT before permeabilizing. This inactivated all surface AChRs. The cells were then permeabilized, fixed, and allowed to react with HRP-labeled αBGT. Without prebinding with αBGT 45% of the coated vesicles were labeled while with prebinding 38% of the coated vesicles were labeled. This finding indicates that most of the coated vesicles which contain AChRs are not endocytotic. On the other hand a 2-hour treatment with puromycin (a blocker of protein synthesis) caused a 3-fold decrease in the number of AChR containing coated vesicles. Since it takes at least 3 hours for newly synthesized AChRs to reach the cell surface (168) this experiment also indicates the exocytotic nature of the receptor containing coated vesicles.

Cultured myotubes also synthesize and secrete acetylcholinesterase (AChE) with the same kinetics and susceptibility to inhibitors as the synthesis and transport of AChRs (168). This finding led Rotundo and Fambrough to postulate that both proteins are transported by the identical vesicular pathway. Recently Benson and Fine (18) have obtained cytochemical evidence that coated vesicles in cultured myotubes contain acetylcholinesterase using the Karnovsky and Roots AChE staining method which produces an iron precipitate visible in the EM (170). Many of the stained coated vesicles were shown to be exocytotic by first treating the cells with diisopropyl fluorophosphate (DFP), a potent inhibitor of active serine proteases and esterases which covalently binds and inactivates all intra- and extracellular AChE. Afterward the cells were washed with fresh medium and allowed to incubate for an additional 2.5 hours, not enough time for any newly synthesized

AChE to reach the cell surface. After staining, many intracellular coated vesicles were again shown to contain AChE. These were shown to be destined for secretion by demonstrating that the above described treatment followed by a further 3-hour incubation with cyclohexamide which blocks protein synthesis led to the disappearance of AChE containing coated vesicles. These findings are confirmed by experiments on coated vesicles isolated from embryonic chick muscle which indicate by biochemical as well as histochemical methods that AChE and AChRs are contained inside of coated vesicles (171, 172).

A recent study (173) suggests that coated vesicles are involved in the transport of newly synthesized neuronal proteins which have traversed the Golgi and are selected for rapid axonal transport.

A recent experiment (172,174) suggests that an individual coated vesicle contains both AChRs and AChE. The reaction product of the Karnovsky–Roots AChE reaction is a very dense iron containing precipitate. Therefore coated vesicles that contain AChE can be separated from those that do not by a Ficoll, sucrose, D_2O gradient. Figure 6 demonstrates this separation. When this experiment is performed on chick muscle coated vesicles, 50% of the coated vesicles density shift after treatment with the AChE substrate. About 90% of the acetylcholine receptors associated with the coated vesicles also density shift after AChE reaction. These findings indicate that a single coated vesicle carries both AChE and AChR molecules, consistent with the postulate that both follow the same vesicular pathway.

Coated vesicles have recently been isolated from DFP-treated cells allowed to recover for 2.5 hours (174). When the density shift experiment is performed with these coated vesicles 50% the protein and over 90% of the AChRs shift. This experiment confirms the fact that both AChE and AChR molecules can be transported in a single coated vesicle during intracellular transport preceding exocytosis. It also suggests that there may be coupled transport of the two proteins.

The fact that over 50% of the isolated coated vesicles contain AChE and are probably exocytotic may be due to any of several reasons. One is that the isolation procedure may have selected for exocytotic coated vesicles. A second and potentially more significant explanation is that in developing muscle cells many of the coated vesicles are in fact engaged in exocytosis. Since these cells are essentially factories for making contractile proteins and are hence rapidly increasing in volume, they need to increase their surface area considerably. If, as seems reasonable, membrane transported via coated vesicles represents a major fraction of the total membrane to be added to or subtracted from the plasma membrane, then the coated vesicle-mediated flow of membrane into the cell surface must be greater than the flow out of the membrane for a cell which is expanding the surface area. Bursztajn and Fischbach (169) have evidence consistent with this explanation. After incubation of a mixed population of cultured

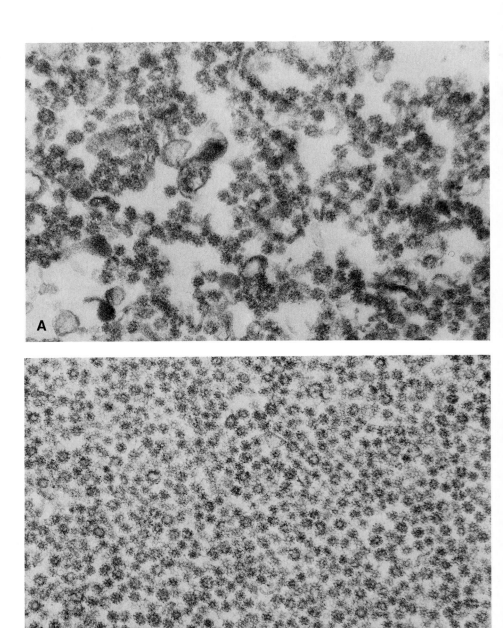

FIG. 6. Thin section electron micrograph showing coated vesicles from embryonic chick skeletal muscle which have been reacted with the Karnovsky–Roots reaction for AChE and centrifuged on a D_2O, Ficoll, sucrose step gradient. (A) Coated vesicles which sediment faster all have darkened lumens. (B) Coated vesicles which don't increase sedimentation rate contain lightly stained lumens. From Porter-Jordan et al. (174) with permission.

muscle cells and fibroblasts with HRP for 1 hour, they observe that the fibroblasts are filled with HRP, as detected histochemically, while the muscle cells contain very little. This indicates that skeletal muscle cells are very inactive in endocytosis, consistent with the possibility that more membrane is being added than removed.

Two other studies are also relevant to this discussion. The work of Heuser and Reese (140) has demonstrated that in the presynaptic neuromuscular junction, coated vesicles are a transient intermediate, seen in large numbers only after a massive synaptic vesicle fusion with the presynaptic plasma membrane, resulting in a large increase in the total surface area. These coated vesicles are definitely endocytotic since they are labeled with HRP which was added to the incubation solution. Here, then is a situation in which the coated vesicle pathway is used mainly if not entirely to remove excess plasma membrane, while not being involved in the insertion of plasma membrane.

Another system which suggests that the coated vesicle pathways in cells can be modified to serve the needs of a pathogen, is the work with the vesicular stomatitis virus (VSV) G protein which will be discussed in more detail in the next section. Coated vesicles from VSV-infected cells contain a very high amount of G protein, the ratio being about 1 G per 3 clathrins (19,175). Recent experiments (175) indicate that coated vesicles isolated from infected cells treated with cyclohexamide for a short time to block protein synthesis contain a greatly reduced amount of G protein. This is consistent with the possibility that coated vesicles mediate the transfer of newly synthesized G protein to the plasma membrane rather than in the endocytosis of G. Very recently it has also been shown that VSV infection very rapidly inhibits endocytosis (176). Together these results suggest that VSV infection results in a shift in the coated vesicle traffic to favor exocytosis. In terms of viral multiplication this seems reasonable since once a few virus particles enter the cell through a coated vesicle-mediated process; further endocytosis becomes detrimental to the formation of new virus since it might remove G protein from the cell surface thereby inhibiting G protein-dependent virus budding.

Obviously these data are merely suggestive but together point to the possibility that shifting the coated vesicle traffic pattern in cells may be an important regulator of membrane dynamics.

WHICH STEPS IN TRANSPORT INVOLVE COATED VESICLES?

The portion(s) of the intracellular pathway involving coated vesicles was not directly indicated in any of the above investigations. As mentioned both ER → Golgi and Golgi → plasma membrane transport routes have been suggested to involve CVs. These questions have been addressed in biochemical experiments suggesting the involvement of coated vesicles in the transport of newly synthesized VSV G proteins (12,19). Because VSV virus infection results in a rapid blockage of cellular protein synthesis, the VSV G protein is the only glycoprotein synthesized and can readily be identified in crude cell extracts after short pulses of a radioactive amino acid are given to infected cells.

By isolating coated vesicles from 40 g quantities of CHO cells infected with VSV, G protein could be directly demonstrated to be present in a rather large amount, 1 G to 3 clathrins, in highly enriched coated vesicle populations. Two other virus polypeptides synthesized on free polysomes and hence not transported to the cell surface by a vesicular route were also present, however, in the initial preparations indicating contamination. More recently the same ratio of G to clathrin has been obtained in coated vesicles preparations purified to virtual homogeneity on agarose gels, while the other viral peptides were completely removed from the preparation (175).

The question of which stage(s) of the G transport involved coated vesicles was addressed using pulse labeling of VSV infected cells with [^{35}S]methionine (12,19). These experiments demonstrated that the newly synthesized protein contained in the isolated coated vesicles is a transit form and at least two successive waves of coated vesicle transport are involved. The oligosaccharides of G_1 carried in the early wave of coated vesicles are sensitive to endoglycosidase H causing a detectable shift in electrophoretic mobility. The second wave consists of G_2 transport; the oligosaccharide at this stage is resistant to Endo H digestion. A kinetic analysis allowed the authors to suggest that the initial stage of G protein transport via coated vesicles was between ER and Golgi apparatus while the second wave was between Golgi and plasma membrane. A recent determination of the oligosaccharides contained in coated vesicle associated G protein from VSV infected cells labeled with [^{3}H]mannose for 30 minutes suggests that coated vesicles may be involved in multiple stages of G protein transport (167). Besides the mannose$_9$ and mature oligosaccharides which would be expected to be contained in ER $\rightarrow$ Golgi coated vesicles and Golgi $\rightarrow$ PM coated vesicles, mannose$_{8-4}$ containing units were also found suggesting that perhaps multiple stages of Golgi $\rightarrow$ Golgi coated vesicle-mediated transfer steps are involved. Rothman (163) has recently suggested multiple transfer stages in the Golgi apparatus on theoretical grounds.

Recently, Wehland *et al.* (178) have reported experiments which are not in agreement with the proposed role of coated vesicles in G protein transport. They microinjected anti-clathrin antibodies and found no detectable blocking of G protein transport. Since, however, this antibody injection also failed to inhibit receptor mediated endocytosis, which is accepted as involving coated pits (60), this finding is of doubtful significance. The second piece of evidence is that simultaneous immunocytochemical localization of both clathrin and G protein in presumed Golgi-associated coated regions indicates the G protein is not present in clathrin coated membrane. As a control they demonstrate that an endocytosed

HRP-EGF conjugate can be demonstrated in Golgi-associated coated structures by a use of similar ferritin-conjugated antibody. This experiment, however, was performed with human KB cells while the G protein localization used mouse 3T3 cells reducing its value as an effective control. Obviously, further investigations must be carried out to resolve this conflict.

Very recently, Kinnon and Owen (179) have purified coated vesicles from a cultured human B lymphocyte cell line. Using Endo H treatment of immunoprecipitated, newly synthesized IgM and HLA-B and DR from coated vesicles isolated from cells pulse labeled with [^{35}S]methionine, they could demonstrate that there was a sequential association of Endo H-sensitive and resistant molecules of the above mentioned proteins with coated vesicles. This work agrees with that of Rothman and Fine (12,19) in indicating that both secretory and integral plasma membrane proteins are carried toward the cell surface in at least two coated vesicle-mediated transfers.

VII. Expression of Coated Vesicle Phenotype

Coated vesicles are found in cells of most mammalian tissues. However, the number of coated vesicles found in different cell types varies. Most cells specialized for protein absorption, e.g., oocytes and enterocytes (7,126) are especially well endowed and contain large numbers of coated vesicles. We are not aware of any evidence that the level of expression varies much in healthy mature cells. The erythrocyte is a good example of a cell where there is evidence that the degree of the expression of coated vesicle phenotype alters during development. The developing cell, the erythroblast, contains coated vesicles which among other functions selectively absorb the protein ferritin (180). However, by the time the mature erythrocyte form is achieved the cell exhibits no coated vesicles or coated pits. In this situation the coated vesicle loss mirrors the behavior of other organelles which are eliminated in this highly specialized cell. There is a set of conditions, however, which can cause the return of coated vesicle expression to mature erythrocytes. Apparently a host cell response occurs when erythrocytes of *Macaca mulatta* are invaded by the intracellular parasite *Plasmodium knowlesi* (181). These authors have shown that host cell membrane coated vesicles return when parasites enter and their level of expression increases up until the schizont stage when the host cell is crammed with parasites.

This type of change is not unprecedented. For example, histiocytes (tissue macrophages) which do not usually form intracellular junctions may produce these organelles in certain pathological states (182). However, the mechanism of providing the information for coated vesicle production in an *anucleate* cell defies explanation by traditional mechanisms based on genome repression.

VIII. Evolution and Coated Vesicles

It is interesting to establish the limits of phylogenetic distribution of phenotypic features which are widespread in cells and organisms. In this respect, much thought has been given to the way in which the acquisition of organelles in evolutionary history has led to the success of increasingly complex life forms (183–185). The distribution of coated vesicles has been reviewed recently by Nevorotin (2); he found coated vesicles in many higher animal groups concluding that they were widely distributed on a species and tissue basis and by Newcomb (186) who has concentrated on their occurrence in plants. Here we wish to focus attention on their occurrence in lower organisms.

A search through ultrastructural papers describing lower organisms shows that coated vesicles are well represented (Table I). The following conclusions can be drawn.

Organelles which appear to be coated vesicles are present in organisms from plant, fungal, and animal kingdoms. These are the three major eukaryotic kingdoms defined by Leedale (200). The finding that coated vesicles are present in myxomycetes and oomycetes (Table I) and their presence in Basidiomycetes (201,202) shows that they were widespread amongst fungi. The acquisition of coated vesicle pathways may be a major reason for the success of the fungi as a group. We have not found evidence for the presence of coated vesicles in the prokaryotic kingdom, the Monera.

Coated vesicles are found, however, in the majority of the lower organisms

TABLE I
COATED VESICLES IN PROTISTS

Kingdom	Group	Organism	Reference
Animalia	Sarcodina	*Acanthocystis erinaceoides*	187
	Ciliophora	*Tetrahymena thermophila*	188
	Zoomastigina	*Trypanosoma raiae*	189
	Porifera	*Spongilla lacustris*	190
Fungi	Myxomycetes	*Dictyostelium discoideum*	191
	Oomycetes	*Saprolegnia ferax*	192
Plantae	Euglenophyta	*Euglena* sp. and *Colacium* sp.	193
	Dinophyta	*Noctiluca* and *Blastodinum*	194
	Cryptophyta	Unidentified species	195
	Chlorophyta	*Pyramimonas parkeae*	196
	Chrysophyta	*Mellomonas caudata*	197
	Baccilariophyceae	Unidentified species symbiotic to *Amphistegina lessonii*	198
	Xanthophyta	Botrydium granulatum	199

defined as Protists (lower eukaryotes). Failure to find evidence among two of the groups may in one case (Chytridomycetes) be owing to the rarity of structural studies of sufficient resolution and in the other (sporozoa) owing to the loss of features so typical of all parasitic forms' evolution.

The evolutionary debut of coated vesicles may therefore coincide with that period in evolutionary history when the transition of certain organisms from prokaryotic form to eukaryote form was in progress. In geological terms this is the late precambrian period. Conversely we may propose that the presence of coated vesicles is another characteristic difference between prokaryotes and eukaryotes (187).

The serial endosymbiotic theory of the evolution of eukaryotes suggests that some organelles commonly found in present day eukaryotes were originally separate organisms which entered the primitive eukaryote. This initially casual symbiotic relationship is thought to have evolved to a state of mutual dependence. The theory is based on the facts that certain bacteria have many similarities to mitochondria, that blue-green bacteria (cyanophyta) have much in common with red algal chloroplasts and that spirochaetes are similar in many ways to cilia. In other words, it explains the complexity of the eukaryotic state by proposing a polyphyletic origin.

A list of 9 points of structural and functional identity between coated vesicles and budding viruses has recently been used to draw attention to their similarities and the possibility that viruses and coated vesicles are related in an evolutionary sense has been discussed (187).

IX. Perspectives and Directions

We can look back with some satisfaction at the number of problems related to coated vesicle biology which have been resolved since 1975. Coated vesicles are demonstrably important mediators of (1) receptor-mediated endocytosis of protein, (2) transepithelial transport, (3) intracellular sorting of proteins, (4) secretion of proteins, (5) membrane biogenesis, and (6) membrane recycling.

These fundamental biological processes are of interest intrinsically, but attention, as we have shown, is now drawn to the more applied areas of pathophysiology. We can cite the role of coated vesicle in (1) familial hypercholesterolemia, (2) toxin action, (3) down regulation of hormone receptors, (4) virus infection, and (5) virus replication as fields of useful recent advances. In an era of biotechnology, it seems possible that there will shortly be applications at least of the principles that the cell uses for sorting and concentrating peptide containing molecules.

The molecular jigsaw of the coated vesicle is almost but tantalizingly not quite complete. The major components are identified but their key properties and functional interaction are yet to be fully described. We need the answers to many questions. For example, is clathrin directly linked to receptors or not? Where and how does the motile force necessary to induce membrane curvature arise? Is receptor occupancy a sufficient or even necessary condition for triggering endocytosis? Precisely how are clathrin and receptors recycled to the surface membrane?

Considering the events occurring when a clathrin lattice forms there are important problems to discuss. First, Heuser shows the submembrane hexagonal lattice of clathrin growing by accretion at its edges rather in the manner predicted (123). However, Crowther and Pearse (57) have shown how simple mutual relative rotation of triskelions may convert six-sided to five-sided figures and thus generate a curved surface in the lattice. Are the dislocations in the lattice observed by Heuser caused by triskelion rotation? Given that there would appear to be components of at least four clathrin molecules along one edge of any lattice polygon would any one triskelion in an extended hexagonal lattice be free to rotate? The probability is that formation of pentagons within a preexisting lattice of hexagons would necessitate the loss of elements of the lattice whether or not the transformation was mediated by the rotation of triskelions. In this respect the beautiful images of Heuser's dislocations (4) require careful study. Some of the heptagons he describes (Fig. 4) have lattice bar-like extensions projecting into their interior. Are these a reflection of damage to or fragility of the lattice or are they owing to the dynamic events supporting the development of curvature. Another related question put colloquially is, how do triskelions cross their legs? The nature of rearrangement of lattice components must be the more complicated if observations of beading in polygon edges prove to derive from an intertwined helical relationship between laterally associated components of triskelions within the lattice.

The question of lattice morphogenesis has been the subject of a mini-review (204). The authors argue that a number of observations made on *in vitro* coated vesicle lattice reconstruction do not support Heuser's interpretation of lattice generation *in vivo* (4). The extended lattices of clathrin seen on macrophage phagosomes ingesting latex and on substrate attached membrane in HeLa cells (144) are probably highly constrained lattices and it may be that Heuser's images relate more directly to structures of these or similar types. They may be subtly different from a coated pit in the process of giving rise to a coated vesicle.

Some of the objections cited echo those made earlier (123). On the grounds of efficiency it is unlikely that a curved lattice forms by loss of a 60° sector of prefabricated hexagonal lattice, but important new evidence is available now about the natural pucker of the triskelion which is probably the "physiological subunit." It can also be seen how triskelions polymerizing in unconstrained conditions give rise to "favored configurations" such as a barrel-like form (55).

The important steps in the process of self-assembly of the coated vesicle are currently being studied intensively. It appears from the evidence that we have reviewed that under physiological conditions the vesicle form with lattice associated is a favorable state and one to which the separated components will return in the laboratory when conditions of salt concentration and pH are appropriate. In view of this the observations that show coated vesicles breaking open to fuse with the membrane of the cell surface during exocytosis or to fuse with another intracellular compartment deserve an explanation. Presumably this is an active process. It may therefore depend upon interesting enzymatic functions. In this regard the recent reports by Rothman's group (71,72) that coated vesicles can be uncoated by a partially purified enzyme in an ATP-dependent process appears to have considerable physiological significance. Presumably vesicle fusions must be closely regulated since inappropriate fusions would seem to be rare if they exist at all. We may speculate that this guidance depends on cytoskeleton-associated transport. In this regard recent evidence for pinocytic vesicle cytoskeletal interactions will hopefully lead to a deeper understanding of the guidance mechanism (20).

Extrapolating from the course of current research we are probably shortly to witness an increase in the number of ligands demonstrated as being taken up or routed by coated vesicles. There could follow an interesting period during which the comparative molecular biology of the uptake of bodily constituents such as serum proteins leads to useful generalizations concerning nutrition and regulation of cells with differing requirements in the same body.

The role of the receptor in coated vesicle-mediated endocytosis has not received as much attention as it warrants. Early indications are that these are glycoproteins with lectin binding characteristics atypical of the cell surface as a whole and that glycolipids are not strong candidates but much work is needed to establish these points and to delineate the characteristic features of a coated vesicle receptor. The geometry of the coated pit is surely of importance to the receptor–ligand interaction. The ligand released from its receptor into a pit during the usual random kinetic movements of particles of molecular size would be constrained to remain in the vicinity of the other receptors in the pit. It would be therefore more likely to bind again sooner to a receptor than would a ligand released from a receptor present at low density on a flat or convex region of cell surface. The notion of cooperativity of receptors needs reexamination in the light of the morphological evidence available to us now about the location and mobility of receptors.

The major drawback of the coated vesicle from a morphologists viewpoint is that it occurs in a variety of sizes and shapes. Many of the image enhancement technologies which have proven so powerful in the elucidation of regular and crystalline structures have not therefore been of much assistance. The ironic probability is that advances in our understanding of this structurally appealing organelle are going mainly to come from biochemical and immunological studies.

The question of whether there is soluble clathrin in the cell remains to be solved. If it is, is it present in triskelion form? Even the limits of distribution of clathrin associated with the inner leaflet of the phospholipid bilayer are presently unclear. Are triskelions, for example, equally partitioned between the apical and basolateral surfaces of absorptive epithelial cells? Are they ever associated with microvillar as opposed to intermicrovillar plasma membrane?

In conclusion we have to reemphasize the far reaching importance of this ubiquitous eukaryotic organelle. It is a sorting and very probably a concentrating organelle. One of the most crucial capabilities of living systems is their ability to survive in an environment which is on average more dilute than the cell in the molecules which are the components of the cell. The coated vesicle appears to be one of the vital instruments which allows this survival.

References

1. Bessis, M., and Breton Gorius, J. (1957). *J. Biophys. Biochem. Cytol.* **3**, 503–505.
2. Nevorotin, A. J. (1980). *In* "Coated Vesicles" (C. D. Ockleford and A. Whyte, eds.), pp. 25–54. Cambridge Univ. Press, London and New York.
3. Kanaseki, T., and Kadota, K. (1969). *J. Cell Biol.* **42**, 202–220.
4. Heuser, J. (1980). *J. Cell Biol.* **84**, 560–583.
5. Scheide, O., San Lin, R., Grellert, E., Galey, J., and Mead, J. (1969). *Physiol. Chem. Phys.* **1**, 141–163.
6. Palay, S. (1963). *J. Cell Biol.* **19**, 89A–90A.
7. Roth, T., and Porter, K. (1964). *J. Cell Biol.* **20**, 313–332.
8. Kartenbeck, J. (1980). *In* "Coated Vesicles" (C. D. Ockleford and A. Whyte, eds.), pp. 243–253. Cambridge Univ. Press, London and New York.
9. Holtzman, E., Novikoff, A., and Villaverde, H. (1967). *J. Cell Biol.* **33**, 419–436.
10. Friend, D., and Farquhar, M. (1967). *J. Cell Biol.* **35**, 357–376.
11. Pearse, B. M. F. (1975). *J. Mol. Biol.* **97**, 93–98.
12. Rothman, J. E., Bursztyn-Pettegrew, H., and Fine, R. E. (1980). *J. Cell Biol.* **86**, 162–171.
13. Pilch, P., Shia, M., Benson, R., and Fine, R. (1983). *J. Cell Biol.* **96**, 133–138.
14. Woods, J. N., Woodward, M., and Roth, T. (1978). *J. Cell Sci.* **30**, 87–97.
15. Pearse, B. M. F. (1976). *Proc. Natl. Acad. Sci. U.S.A.* **75**, 4394–4398.
16. Mello, R. J., Brown, M. S., Goldstein, J. L., and Anderson, R. G. (1980). *Cell* **20**, 829–837.
17. Ockleford, C. D., Whyte, A., and Bowyer, D. (1977). *Cell Biol. Int. Rep.* **3**, 717–724.
18. Benson, R. J., Porter-Jordan, K., and Fine, R. E. (1984). *J. Cell Biol.* submitted.
19. Rothman, J., and Fine, R. E. (1980). *Proc. Natl. Acad. Sci. U.S.A.* **77**, 780–784.
20. Keen, J. H., Willingham, M. C., and Pastan, I. (1979). *Cell* **16**, 303–312.
21. Schook, W., Puszkin, S., Bloom, W., Ores, C., and Kochwa, S. (1980). *Proc. Natl. Acad. Sci. U.S.A.* **76**, 116–120.
22. Booth, A., and Wilson, M. (1981). *Biochem. J.* **196**, 355–362.
23. Pearse, B.M.F. (1982). *Proc. Natl. Acad. Sci. U.S.A.* **79**, 451–445.
24. Nandi, P. K., Irace, G., Van Jaarsveld, P. V., Lippoldt, R. E., and Edelhach, H. (1983). *Proc. Natl. Acad. Sci. U.S.A.* **79**, 5881–5885.
25. Fine, R., and Squicciarini, J. (1984). In preparation.
26. Pauloin, A., Bernier, I., and Jolle's P. (1982). *Nature (London)* **298**, 574–576.
27. Pfeiffer, S. R., and Kelley, R. (1981). *J. Cell Biol.* **91**, 385–391.

28. Mueller, S., and Branton, D. (1982). *J. Cell Biol.* **95**, 389a.
29. Rubenstein, J.R.R., Fine, R., Luskey, B.. and Rothman, J. (1981). *J. Cell Biol.* **89**, 357–361.
30. Merisko, E., Farquhar, M., and Palade, G. (1982). *J. Cell Biol.* **92**, 846–857.
31. Robinson, M. S. (1982). Ph.D. thesis, Harvard University.
32. Mersey, B. G., Fowke, L. C.. Constable, F., and Newcombe, E. H. (1982). *Exp. Cell Res.* **141**, 459–463.
33. Fine, R. E., Blitz, A. L., and Sack, D. (1978). *FEBS Lett.* **94**, 59–61.
34. Blitz, A. L., Fine, R. E., and Toseli, P., (1977). *J. Cell Biol.* **75**, 135–147.
35. Unanue, E., Ungewickell, E., and Branton, D. (1981). *Cell* **26**, 439–445.
36. Pearse, B.M.F., and Bretscher, M. S. (1981). *Annu. Rev. Biochem.* **50**, 85–101.
37. Robinson, M. S. (1978). *Biol. Bull.* **155**, 263.
38. Sack, D., Fine, R., and Blitz, A. (1978). *Fed. Proc. Fed. Am. Soc. Exp. Biol.* **37**, 278A.
39. Pfeffer, S., Drubin, D., and Kelly, R. (1983). *J. Cell Biol.* **97**, 40–48.
40. Fine, R., Shia, M., Campbell, C., Squicciarini, J., and Pilch, P. (1983). *Fed. Proc. Fed. Am. Soc. Exp. Biol.* **42**, 2249a.
41. Campbell, C., Shia, M., Squicciarini, J., Pilch, P., and Fine, R. (1984). *Biochemistry,* in press.
42. Linden, C., Dedman, J., Chafouleas, J., Means, A., and Roth, T. (1981). *Proc. Natl. Acad. Sci. U.S.A.* **79**, 308–312.
42a. Linden, C. (1982). *Biochem. Biophys. Res. Commun.* **109**, 186–193.
43. Lisanti, M. P., Shapiro, L., Moskowitz, N., Hua, E., Puszkin, S., and Schook, W. (1982). *Eur. J. Biochem.* **125**, 463–470.
44. Moskowitz, N., Glassman, A., Ores, C., Schook, W., and Puszkin, S. (1983). *J. Neurochem.* **40**, 711–718.
45. Nandi, P. K., Pretorius, H. T., Lippoldt, R. E., Johnson, M. L.. and Edelhoch, H., (1980). *Biochemistry* **19**, 5917–5921.
46. Pearse, B.M.F. (1978). *J. Mol. Biol.* **126**, 803–812.
47. Woodward, M. P., and Roth, T. F. (1978). *Proc. Natl. Acad. Sci. U.S.A.* **75**, 785–798.
48. Ungewickell, E., and Branton, D. (1981). *Nature (London)* **289**, 420–422.
49. Kirchhausen, T., and Harrison, S. C. (1981). *Cell* **23**, 755–761.
50. Lisanti, M. P., Schook, W., Moskowitz, N., Ores, C., and Puszkin, S. (1981). *Biochem. J.* **201**, 297–304.
51. Schmid, S., Matsumoto, A., and Rothman, J. (1982). *Proc. Natl. Acad. Sci. U.S.A.* **79**, 91–95.
52. Ungewickell, E., Unanue, E., and Branton, D. (1982). *Cold Spring Harbor Symp. Quant. Biol.* **46**, 723–731.
53. Kirchhausen, T., Harrison, S., Parham, P., and Brodksy, F. (1983). *Proc. Natl. Acad. Sci. U.S.A.* **80**, 2481–2485.
54. Brodsky, F., Holmes, N., and Parham, P. (1983). *J. Cell Biol.* **96**, 911–914.
55. Crowther, R. A., Finch, J. T., and Pearse, B. M. (1976). *J. Mol. Biol.* **103**, 785–798.
56. Ockleford, C. D., De-Voy, K., and Hall, H. M. (1979). *Cell Biol. Int. Rep.* **3**, 717–724.
57. Crowther, R. A., and Pearse, B. M. (1981). *J. Cell Biol.* **91**, 790–797.
58. Steer, C. J., Klausner, R. D., and Blumenthal, R. (1982). *J. Biol. Chem.* **257**, 8533–8540.
59. Blumenthal, R., Henkart, M., and Steer, C. (1983). *J. Biol. Chem.* **258**, 3409–3415.
60. Anderson, R. G., Goldstein, J., and Brown, M. (1977). *Cell* **15**, 919–933.
61. Clarke, B., and Weigel, P. H. (1982). *J. Cell Biol.* **95**, 442a.
62. Salisbury, J. L., Condeelis, J. S., and Satir, P. (1980). *J. Cell Biol.* **87**, 132–141.
63. Ben-Svi, A., and Abulafia, R. (1982). *Mol. Cell Biol.* **3**, 684–692.
64. Maxfield, F., Willingham, M., Davies, P., and Pastan, I. (1979). *Nature (London)* **277**, 661–663.

65. Davies, P., Davies, D., Levitzki, A., Maxfield, F., Milhaud, P., Willingham, M. C., and Pastan, I. (1980). *Nature (London)* **283**, 162–167.
66. Salisbury, J., Condeelis, J., Maihle, N., and Satir, P. (1981). *Nature (London)* **294**, 163–166.
67. Pastan, I., and Willingham, M. (1981). *Science* **214**, 504–509.
68. Wehland, J., Willingham, M., Dickson, R., and Pastan, I. (1981). *Cell* **25**, 105–119.
69. Fan, J., Carpentier, J., Gorden, P., Van Obberghen, E., Blackett, N., Grunfeld, C., and Orci, L. (1982). *Proc. Natl. Acad. Sci. U.S.A.* **79**, 7788–7791.
70. Petersen, O. W.. and Van Deurs, B. (1983). *J. Cell Biol.* **96**, 277–281.
71. Patzer, E., Schlossman, D., and Rothman, J. (1982). *J. Cell Biol.* **93**, 230–236.
72. Schmid, S., Schlossman, D., Braell, W., and Rothman, J. (1983). *Fed. Proc. Fed. Am. Soc. Exp. Biol.* **42**, 1803a.
73. Bliel, J., and Bretscher, M. (1982). *EMBO J.* **1**, 351–355.
74. Wall, D. A., Wilson, G., and Hubbard, A. L. (1981). *Cell* **21**, 79–93.
75. Schwartz, A., Fridovich, S., and Lodish, H. (1982). *J. Biol. Chem.* **257**, 4230–4237.
76. Maxfield, F. R., Schlessinger, J., Shecter, J., Pastan, I., and Willingham, M. (1978). *Cell* **14**, 805–810.
77. Willingham, M., Maxfield, F., and Pastan, I. (1979). *J. Cell Biol.* **82**, 614–625.
78. Carpentier, J. L., Gorden, P., Anderson, R., Goldstein, J., Brown, M., Cohen, S., and Orci, L. (1982). *Cell Biol.* **95**, 73–77.
79. Connolly, J., Greene, L., Viscarello, R., and Riley, W. (1979). *J. Cell Biol.* **82**, 820–827.
80. Connolly, J., Green, S., and Greene, L. (1981). *J. Cell Biol.* **91**, 597a.
81. Anderson, R. G., Goldstein, J., and Brown, M. S. (1976). *Nature (London)* **270**, 695–699.
82. Bretscher, M. S., Thomson, J. N., and Pearse, B.M.F. (1981). *Proc. Natl. Acad. Sci. U.S.A.* **77**, 4159–4163.
82a. Bergman, J. F., and Carvey, D. H. (1982). *J. Cell Biochem.* **20**, 247–258.
83. Robinson, M., Rhodes, J., and Albertini, D. (1983). *J. Cell. Physiol.,* **117**, 43–50.
84. Orci, L., Carpentier, J., Perrelet, A., Anderson, R. G., Goldstein, J., and Brown, M. S. (1978). *Exp. Cell Res.* **113**, 1–13.
85. Geuze, H., Slot, J.. Strous, G., Lodish, H., and Schwartz, A. (1983). *Cell* **32**, 277–287.
86. Anderson, R. G., Brown, M. S., Beisiegal, V., and Goldstein, J. (1982). *J. Cell Biol.* **93**, 523–531.
87. Steer, C. J., and Ashwell, G. (1980). *J. Biol. Chem.* **255**, 3008–3013.
88. Fehlman, M., Carpentier, J., LeCam, A., Thamm, P., Saunders, D., Brandenberg, D., Orci, L., and Freychet, P. (1982). *J. Cell Biol.* **93**, 82–87.
89. Ryser, H., Drummond, I., and Shen, W. (1982). *J. Cell. Physiol.* **113**, 167–178.
90. Haigler, H., McKanna, J., and Cohen, S. (1979). *J. Cell Biol.* **83**, 82–90.
91. Fine, R. E., Goldenberg, R., Sorrentino, J., and Herschman, H. R. (1981). *J. Supramol. Struct.* **15**, 235–251.
92. Das, M., and Fox, C. F. (1978). *Proc. Natl. Acad. Sci. U.S.A.* **75**, 2644–2648.
93. Gavin, J. R., Roth, J., Neville, D., DeMeyts, P., and Buell, D. (1974). *Proc. Natl. Acad. Sci. U.S.A.* **71**, 84–88.
94. Marsh, M., and Helenius, A. (1981). *J. Mol. Biol.* **142**, 439–454.
95. Draper, R. K., and Simon, M. I. (1980). *J. Cell Biol.* **87**, 849–854.
96. Ray, B., and Wu, H. C. (1981). *Mol. Cell Biol.* **1**, 557–559.
97. Sandwig, K., and Olsnes, S. (1980). *J. Cell Biol.* **87**, 828–832.
98. Gerstein, A., Morris, R., and Saelinger, C. (1982). *J. Cell Biol.* **99**, 418a.
99. Motesano, R., Roth, J., Robert, A., and Orci, L. (1982). *Nature (London)* **296**, 651–653.
100. Okhuma, S., and Poole, B. (1978). *Proc. Natl. Acad. Sci. U.S.A.* **75**, 3327–3331.
101. Basu, S., Goldstein, J., Anderson, R., and Brown, M. (1981). *Cell* **24**, 493–502.
102. Marsh, M., Wellstead, J., Kern, H., Harms, E., and Helenius, A. (1982). *Proc. Natl. Acad. Sci. U.S.A.* **79**, 5297–5301.

103. King, A., Hernea-Davis, L., and Cautrecasas, P. (1980). *Proc. Natl. Acad. Sci. U.S.A.* **77,** 3283–3287.
104. Leppla, S., Dorland, R., and Middlebrook, J. (1980). *J. Biol. Chem.* **255,** 2247–2250.
105. Maxfield, F., Willingham, M., Davies, P., and Pastan, I. (1979). *Nature (London)* **277,** 661–663.
106. Libby, P., Bursztajn, S., and Goldberg, A. (1980). *Cell* **19,** 481–491.
107. Tycko, B., and Maxfield, F. (1982). *Cell* **78,** 643–651.
108. Marsh, M., Wellsteed, J., Bolzau, E., and Helenius, A. (1982). *J. Cell Biol.* **99,** 418a.
109. Dunn, W., Hubbard, A., and Aronson, N. (1980). *J. Biol. Chem.* **255,** 5971–5978.
110. Brooks, R., Merion, M., Sly, W., and Canonico, P. (1982). *J. Cell Biol.* **95,** 439a.
111. Quintart, J., Courtoy, P., and Baudhuin, P. (1982). *J. Cell Biol.* **95,** 424a.
112. Galloway, C., Dean, G., Marsh, M., Rudnick, G., and Mellman, I. (1983). *Proc. Natl. Acad. Sci. U.S.A.* **80,** 3334–3338.
113. Forgac, M., Cantley, L., Wiedenmann, B., Altstiel, L., and Branton, D. (1983). *Proc. Natl. Acad. Sci. U.S.A.* **80,** 1300–1303.
114. Stone, D., Xie, X. S., and Racker, E. (1983). *J. Biol. Chem.* **258,** 4059–4062.
115. Schneider, D. (1981). *J. Biol. Chem.* **256,** 3858–3864.
116. Mellman, I., and Plutner, H. (1982). *J. Cell Biol.* **95,** 434a.
117. Larkin, J., Brown, M., Goldstein, J., and Anderson, R. (1983). *Cell* **33,** 273–285.
118. Debanne, M., Evans, W., Flint, N., and Regoeczi, E. (1982). *Nature (London)* **298,** 398–400.
119. Blumenthal, R., Klausner, R., and Weinstein, J. (1980). *Nature (London)* **288,** 333–338.
120. Dunn, W., and Hubbard, A. (1982). *J. Cell Biol.* **95,** 425a.
121. Pilch, P., Shia, H., Benson, R. J., and Fine, R. (1983). *J. Cell Biol.* **96,** 133–138.
122. Ockleford, C. D., and Whyte, A. (1977). *J. Cell Sci.* **25,** 293–312.
123. Ockleford, C. D. (1976). *J. Cell Sci.* **21,** 83–91.
124. Dearden, L., and Ockleford, C. D. (1983). *In* "The Biology of Trophoblast" (Y. W. Loke and A. Whyte, eds.), Vol. I, pp. 69–109. North-Holland Publ., Amsterdam.
125. Ockleford, C. D., and Clint, J. M. (1980). *Placenta* **1,** 91–111.
126. Rodewald, R. (1973). *J. Cell Biol.* **58,** 198–211.
127. Abrahamson, D. R., and Rodewald, R. (1981). *J. Cell Biol.* **91,** 270–280.
128. Huxham, M., and Beck, F. (1981). *Cell Biol. Int. Rep.* **5,** 1073–1081.
129. Wada, H., Hass, P. E., and Sussman, H. (1979). *J. Biol. Chem.* **254,** 12629–12635.
130. Seligman, P., Schleicher, R., and Allen, R. (1979). *J. Biol. Chem.* **254,** 9945–9946.
131. Karin, M., and Mintz, B. (1981). *J. Biol. Chem.* **256,** 3245–3252.
132. Dautry-Varsat, A., Ciecanover, A., and Lodish, H. (1983). *Proc. Natl. Acad. Sci. U.S.A.* **80,** 2258–2262.
133. Klausner, R., Ashwell, G., Vanrenswoude, J., Hartford, J., and Bridges, K. (1983). *Proc. Natl. Acad. Sci. U.S.A.* **80,** 2263–2266.
134. Regoeczi, E., Taylor, P., Debanne, M., and Charlwood, P. (1982). *Proc. Natl. Acad. Sci. U.S.A.* **79,** 2226–2230.
135. Herzog, V. (1981). *Trends Biochem. Sci.* **6,** 319–322.
136. Farquhar, M. G. (1978). *J. Cell Biol.* **78,** 35–42.
137. Aggeler, J., and Werb, Z. (1982). *J. Cell Biol.* **94,** 613–623.
138. Takemura, R., Aggeler, J., and Web, Z. (1982). *J. Cell Biol.* **95,** 423a.
139. Maupin, P., and Pollard, T. (1983). *J. Cell Biol.* **96,** 51–62.
140. Heuser, J., and Reese, T. (1973). *J. Cell Biol.* **57,** 315–344.
141. Heuser, J., Reese, T., and Landis, D. (1975). *Cold Spring Harbor Symp. Quant. Biol.* **40,** 17–24.
142. Heuser, J. (1978). *In* "Transport of Macromolecules in Cellular Systems" (S. C. Silverstein, ed.), pp. 445–464. Dahlen Konferenzer, Berlin.
143. Geuze, H., and Kramer, M. (1974). *Cell Tissue Res.* **156,** 1–20.

144. Pollard, H., Pazoles, C., Creutz, C., and Zinder, O. (1979). *Int. Rev. Cytol.* **58,** 159–197.
144a. Fisher, G. W., and Rebhun, L. (1983). *Dev. Biol.* **99,** 456–472.
145. Bretscher, M. (1982). *Proc. Natl. Acad. Sci. U.S.A.* **80,** 454–458.
146. Bretscher, M., and Thomson, N. (1983). *EMBO J.* **2,** 599–603.
147. Bretscher, M. (1976). *Nature (London)* **260,** 21–23.
148. Steinman, R., and Cohn, Z. (1974). *J. Cell Biol.* **63,** 949–969.
149. Palay, S. (1963). *J. Cell Biol.* **19,** 89A–90A.
150. Decker, R. S. (1974). *J. Cell Biol.* **61,** 599–612.
151. Natowicz, M., Chi, M., Lowry, O., and Sly, W. (1979). *Proc. Natl. Acad. Sci. U.S.A.* **76,** 4322–4326.
152. Neufeld, E. (1977). *Trends Biochem. Sci.* **2,** 25–26.
153. Rome, L., Weissman, B., and Neufeld, E. (1979). *Proc. Natl. Acad. Sci. U.S.A.* **76,** 2331–2334.
154. Fischer, H., Gonzalez-Noriega, A., Sly, W., and Morre, D. (1980). *J. Biol. Chem.* **255,** 9608–9615.
155. Willingham, M., Pastan, I., Sahagian, G., Jourdian, G., and Neufeld, E. (1981). *Proc. Natl. Acad. Sci. U.S.A.* **78,** 6967–6971.
156. Sahagian, G., Distler, J., and Jourdian, G. (1981). *Proc. Natl. Acad. Sci. U.S.A.* **78,** 4289–4293.
157. Steiner, A. W., and Rome, L. (1982). *Arch. Biochem. Biophys.* **214,** 681–687.
158. Campbell, C., Fine, R., Squicciarini, J., and Rome, L. (1983). *J. Biol. Chem.* **258,** 2628–2633.
159. Campbell, C., and Rome, L. (1983). *J. Biol. Chem.* **258,** 13347–13352.
160. Palade, G. E. (1975). *Science* **189,** 347–358.
161. Hunt, L., and Summers, D. (1976). *J. Virol.* **20,** 646–657.
162. Bergmann, J., Tokuyasu, K., and Singer, S. (1981). *Proc. Natl. Acad. Sci. U.S.A.* **78,** 1746–1750.
163. Rothman, J. (1981). *Science* **213,** 1212–1220.
164. Jamieson, J., and Palade, G. (1971). *J. Cell Biol.* **50,** 135–158.
165. Franke, W., Luder, M., Kartenbeck, J., Zerban, H., and Keenen, T. (1976). *J. Cell Biol.* **69,** 173–195.
166. Ehrenreich, J., Bergeron, J., Siekewitz, P., and Palade, G. (1973). *J. Cell Biol.* **59,** 45–72.
167. Palade, G., and Fletcher, M. (1977). *J. Cell Biol.* **75,** 371a.
168. Rotundo, R., and Fambrough, D. (1980). *Cell* **22,** 595–602.
169. Bursztajn, S., and Fischbach, G. (1984). *J. Cell Biol.* **98,** 498–506.
170. Karnovsky, M. J., and Roots, L. (1964). *J. Histochem. Cytochem.* **12,** 219–221.
171. Porter-Jordan, K., Johnson, R., and Fine, R. (1981). *J. Cell Biol.* **91,** 393A.
172. Porter-Jordan, K., Benson, R. J., Buoniconti, P., and Fine, R. (1984). *J. Cell Biol.,* submitted.
173. Stone, G. C., Hammerschlag, R., and Bobinski, J. A. (1982). *J. Cell Biol.* **95,** 405a.
174. Porter-Jordan, K., Johnson, R.. and Fine, R. (1982). *J. Cell Biol.* **95,** 465a.
175. Rubenstein, J., Fine, R., and Rothman, J. (1984). In preparation.
176. Wilcox, D., Whitaker-Dowling, P., Youngner, J., and Widnell, C. (1982). *J. Cell Biol.* **95,** 445a.
177. Snyder, M., Fine, R., and Rothman, J. (1983). Unpublished results.
178. Wehland, J., Willingham, M., Gallo, M., and Pastan, I. (1982). *Cell* **28,** 831–842.
179. Kinnon, C., and Owen, M. (1983). *J. Biol. Chem.* **258,** 8470–8476.
180. Fawcett, D. W. (1966). *In* "An Atlas of Fine Structure—The Cell," pp. 365–382. Saunders, Philadelphia, Pennsylvania.
181. Wunderlich, F., Stubig, H., and Koniqk, H. (1982). *J. Protozool.* **24,** 49–59.

182. Caputo, R., and Gianotti, F. (1979). *J. Ultrastruct. Res.* **68,** 256–264.
183. Margulis, L. (1970). "Origin of Eukaryotic Cells." Yale Univ. Press, New Haven, Connecticut.
184. Cavalier-Smith, T. (1975). *Nature (London)* **256,** 463–468.
185. Tribe, M. A., Morgan, A. J., and Whittaker, E. (1981). "The Evolution of Eukaryotic Cells." Arnold, London.
186. Newcomb, E. H. (1980). *In* "Coated Vesicles" (C. D. Ockleford and A. Whyte, eds.), pp. 55–69. Cambridge Univ. Press, London and New York.
187. Ockleford, C. D. (1981). *In* "Coated Vesicles in Electron Microscopy of Proteins" (J. R. Harris, ed.), pp. 255–299. Academic Press, New York.
188. Satir, B. H., and Wissig, S. L. (1982). *J. Cell Sci.* **55,** 13–33.
189. Preston, T. M. (1969). *J. Protozool.* **16,** 320–333.
190. Simpson, T. L., Refolo, L. M., and Kaby, M. E. (1979). *J. Morphol.* **159,** 343–354.
191. Favard-Sereno, C., Ludosky, M. A., and Rytor, A. (1981). *J. Cell Sci.* **51,** 63–84.
192. Heath, I. B., and Greenwood, A. D. (1971). *Z. Zellforsch.* **112,** 371–389.
193. Leedale, G. F. (1967). *Annu. Rev. Microbiol.* **21,** 31–48.
194. Soyer, M. O. (1970). *Z. Zellforsch.* **105,** 350–388.
195. Hausmann, K. (1979). *Arch. Protistenk.* **122,** 222–225.
196. Norris, R. E., and Pearson, B. R. (1975). *Arch. Protistenk.* **117,** 192–213.
197. Wujek, D. E., and Kritsiansen, J. (1978). *Arch. Protistenk.* **120,** 213–221.
198. Berthold, V. W-U. (1978). *Arch. Protistenk.* **120,** 16–62.
199. Falk, H. (1969). *J. Cell Biol.* **63,** 167–174.
200. Leedale, G. F. (1974). *Taxon* **23,** 261–270.
201. Grove, S. N., and Bracher, C. E. (1970). *J. Bacteriol.* **104,** 989–1009.
202. Raven, P. H. (1970). *Science* **169,** 641–646.
203. Herman, B., and Albertini, D. (1982). *Cell Motil.* **2,** 583–597.
204. Harrison, S., and Kirchhausen, T. (1983). *Cell* **33,** 650–652.

INTERNATIONAL REVIEW OF CYTOLOGY, VOL. 91

Structure and Function of the Crustacean Larval Salt Gland

FRANK P. CONTE

Department of Zoology, Oregon State University, Corvallis, Oregon

I. Introduction: Osmotic Strategies Utilized by Halobionts

Halobionts are organisms that can live, reproduce, and complete their life cycle in concentrated brines. These organisms can be phylogenetically quite diverse, ranging from the primitive halobacteria and unicellular algae to complex vascular plants and animals. A harsh environmental habitat, such as is found in inland salt lakes and coastal salterns, usually demands that organisms evolve

45

special adaptations to osmotic stress in order to maintain water and electrolyte balance compatible with their life cycle. For halobionts which seek "osmotic conformity," life is sustained in large measure by the fact that the internal osmotic pressure within individual cells is controlled by accumulating a compatible organic solute (osmolyte). Somero and colleagues (Yancey *et al.*, 1982) have cogently argued that halophilic osmoconformers utilize a small family of organic osmolytes thereby achieving a genetically "simpler" and more flexible adaptive mechanism to counter the cyclic water stress than those of the "primitive" halobionts. Halobacteria achieve similar control over intracellular osmotic pressures by genetic manipulation of macromolecules that comprise the subcellular organelles (i.e., ribosomes). Many of the cytoplasmic proteins of halobacteria have undergone extensive amino acid substitutions, primarily in the acidic amino acids. These substitutions enable cytosolic proteins to be more functional in the presence of high intracellular inorganic osmolyte (K^+). In fact, many halobacteria enzymes have an obligatory requirement for K^+ ion to be in the range of 1 M in order for optimal catalytic conditions to exist. It is obvious that genetic evolution through natural selection has taken a long period of time for this system to ascertain which of the many changes in DNA sequences needed to code for these proteins was best in coping with widely fluctuating environmental salinities.

However, for those halobionts that sought mechanisms of "osmotic regulation," life is sustained by the fact that control of intracellular osmotic pressure can be achieved indirectly via regulation of the chemical composition of the extracellular fluid surrounding tissue cells. The mechanism for the control of extracellular fluid appears to be manifested in the microdomains of cell surfaces, especially so for the plasma membrane surrounding individual epithelial cells lining renal and extrarenal organs.

To illustrate my point, I shall briefly focus on two classes of aquatic arthropods, the insects and crustaceans, that are among the most successful animal halobionts.

A. HALOPHILIC INSECTS

The larvae of the saline-water mosquito *Aedes taeniorhynchus* has been extensively studied by Phillips and Bradley (Bradley and Phillips, 1975, 1977a–c; Phillips and Bradley, 1977) and a recent review by Bradley (1983) highlights the functions of the excretory system in this insect. The strategy employed by the mosquito larvae is not to conserve water, which is readily available in their environment, but rather to drink the external water and secrete the ions which contaminate it. The ingested water is absorbed by the mid-portion of the alimentary canal. The ultrastructural properties of the malpighian tubules resemble other insect species (Bradley *et al.*, 1982). Comparison of the functional aspects of the malpighian tubules to the posterior rectal pads has shown that the urine exiting from the malpighian tubules is isosmotic to the hemolymph but high in KCl, organic nutrients, and metabolic wastes. Sometimes the urine is rich in Mg^{2+} and SO_4^{2-} ion depending upon the composition of external saline. This isosmotic filtrate primarily serves to remove waste products and not as a water conservation measure. The major mechanism for water conservation is achieved by other structures such as the anal papillae (Phillips and Meredith, 1969; Meredith and Phillips, 1973a,b) and the rectal salt gland (Asakura, 1970; Bradley and Phillips, 1975) acting as extrarenal organs. The rectal salt gland is capable of actively secreting numerous ions directly into the lumen against a concentration gradient, thereby enriching the ionic content of the luminal fluid. This secretory process diminishes the salt concentration of the hemolymph (Fig. 1A) and brings the ECF into osmotic equilibrium with the intracellular fluids. The specific molecular complexes controlling which ions will be actively transported and at what rate reside in the microdomains of the plasma membrane and are unknown (Phillips, 1982).

B. Halophilic Crustaceans

The adult brine shrimp, *Artemia,* has been extensively studied by Croghan (1958a–e), Smith (1969a,b), and Thuet *et al.* (1968) while Geddes (1975a–c) has investigated a related species, *Parartemia,* regarding osmotic and ionic regulatory properties. The basic strategy employed by the adult shrimp is, in many ways, identical to that found in the larval mosquito. The animal drinks the external medium and the ingested fluid is absorbed from the midgut at a rate equal to fluid loss across the body surface and through the kidney. However, crustaceans being a different class of arthropod do not possess an excretory system comprised of malpighian tubules, rectal, and anal salt glands. Instead the maxillary gland is considered the major renal organ. Like the malpighian tubule, the urine from the maxillary gland is always isosmotic (Potts and Parry, 1964). As a consequence, the inability of the maxillary gland to make a hyperosmotic urine prevents the gland from eliminating the salt load being carried in the hemolymph from bathing tissue cells. Therefore, like the larval mosquito, adult brine shrimp have need of extrarenal organs. In searching various anatomical parts, which might be serving as extrarenal sites, Copeland (1966, 1967), together with Croghan (1958c), established the experimental evidence to support the hypothesis that the leaf-like structures located in the middle portion of individual thoracic swimming appendages contained the ion secreting epithelia (Fig. 2A, B, C, and D). These leaf-like organs provide the adult shrimp with a dual physiological role: first, they can serve as a gill having a thin respiratory surface for gaseous exchange, and second, as an auxillary excretory site for salt extrusion. The gill (branchial) salt gland of the adult brine shrimp could be considered

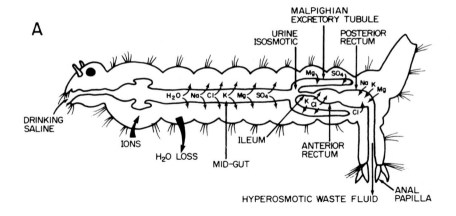

A

MALPIGHIAN
EXCRETORY TUBULE

URINE
ISOSMOTIC

POSTERIOR
RECTUM

Mg SO₄

Na K Mg

H₂O Na Cl K Mg SO₄

K Cl

Cl

DRINKING
SALINE

IONS

H₂O LOSS

MID-GUT

ILEUM

ANTERIOR
RECTUM

HYPEROSMOTIC WASTE FLUID

ANAL
PAPILLA

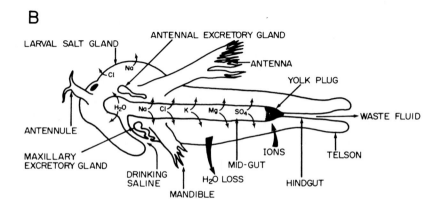

B

LARVAL SALT GLAND

ANTENNAL EXCRETORY GLAND

ANTENNA

YOLK PLUG

Na

Cl

H₂O Na Cl K Mg SO₄

WASTE FLUID

ANTENNULE

MAXILLARY
EXCRETORY GLAND

DRINKING
SALINE

MANDIBLE

H₂O LOSS

MID-GUT

IONS

HINDGUT

TELSON

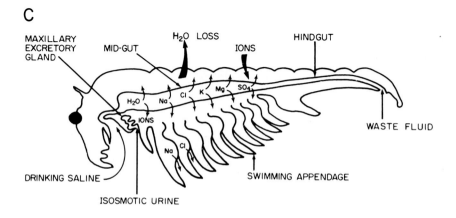

C

MAXILLARY
EXCRETORY
GLAND

MID-GUT

H₂O LOSS

IONS

HINDGUT

H₂O Na Cl K Mg SO₄

IONS

Na Cl

WASTE FLUID

DRINKING SALINE

ISOSMOTIC URINE

SWIMMING APPENDAGE

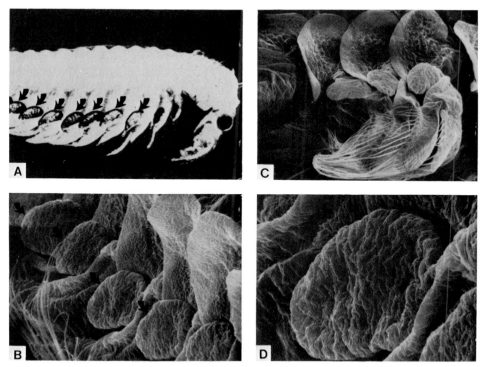

FIG. 2. (A) Adult brine shrimp with AgNO₃-stained gill salt glands (×100); (B) scanning electron photomicrographs of metepipodites of swimming appendage with arrows at salt glands (×200); (C) enlargement epipodite, endopodite, and metepodite structures (×100); (D) solitary gill salt gland (×2400).

as a functional analog to the rectal salt gland of the larval mosquito. Again, it must be stated that specific cellular and molecular mechanisms which can explain the physiological properties of ion transport in gill salt gland epithelial cells are lacking.

Larval brine shrimp, like their adult counterparts, are capable of living in saturated brines. Early investigations on water and electrolyte balance indicated that the naupliar stage possessed an excretory system capable of providing osmotic and ionic regulation (Croghan, 1958b; Conte *et al.*, 1972; Ewing *et al.*,

FIG. 1. (A) Hypoosmotic regulation in halophilic insect (brine mosquito larvae with small anal papillae, malpighian tubule, and rectal gland); (B) hypoosmotic regulation in halophilic crustacean (brine shrimp larvae with antennal gland, maxillary gland, and salt secretory organ located in cephalothoracic region); (C) hypoosmotic regulation in halophilic crustacean (brine shrimp adult with maxillary gland and salt secretory organ located on middle segment of swimming appendage).

1972). Surprisingly, larval and adult brine shrimp exhibit as much morphological divergence in excretory organs as was found to exist between crustaceans and insects.

II. Developmental Pattern in Larval Brine Shrimp

Since osmotic and ionic regulation in the adult brine shrimp is dependent upon these structures, it would enlighten our understanding of their role (or lack of it) in larval brine shrimp if we review the developmental pattern of the excretory glands and alimentary tract. Warren (1938) and Weisz (1947) have studied the various naupliar, metanaupliar, and juvenile stages of brine shrimp and reported on the morphological and developmental pattern of the segmentally arranged excretory glands.

A. EXCRETORY GLANDS

Heath (1924) described in detail the external appearance and morphological development of individual larval stages. For example, a first instar nauplius is depicted in Fig. 3. When the naupliar external morphology is compared to the adult (Fig. 2), one can readily observe the absence of swimming appendages with the attendant loss of gill salt glands. This anatomical paradox poses a series of interesting questions about the physiology of larval brine shrimp. For instance, (1) Does the first instar nauplius possess an excretory system comprised of a functional maxillary gland or are there other excretory glands which operate in its place? (2) Does the first instar nauplius possess a functional alimentary canal which can absorb water and ions from ingested fluid?

1. *Antennal Gland*

Warren (1938) and Weisz (1947) found, and we have confirmed (unpublished observation), that the anlage of the antennal glands are present in the prenaupliar stage (late emergent stages known as E_2 and shown in Fig. 4) as mesenchymal aggregations near the base of the antennae. At hatch, a first-instar nauplius does contain a complete schizocoelic end-sac, but does not have a completed excretory duct. The excretory duct, which opens to the outside near the posterior portion of the protopodite of the antenna, is finished some 10–20 hours following the first-instar stage. Therefore, the antennal gland is fully developed (functional) only at the second-instar stage. Interestingly, by the eleventh-instar the antennal gland has completely disappeared making it a transient larval structure.

2. *Maxillary Gland*

The maxillary gland anlage is absent in the first and second instar nauplius. Formation of the schizocoelic end-sac begins in the third instar but the gland is

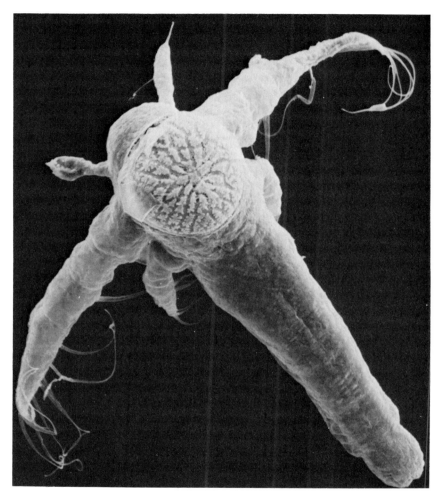

FIG. 3. Scanning electron micrograph of a first instar nauplius larva of the brine shrimp, *Artemia salina*. The salt gland is the oval structure located atop the dorsal surface of the cephalothorax. It is possible to see individual cells with scalloped boundaries. Note the absence of legs and gill leaflets from the long slender abdomen (×200). Taken from Hootman *et al.* (1972).

not complete until the sixth- or seventh-instar metanauplius (Warren, 1938). The maxillary gland cannot be the naupliar renal organ due to the delay in its formation, nor is it a transient larval structure. Apparently, it is destined only to become the adult renal organ.

3. *Mandibular Gland*

The mandibular gland anlage is also absent in the first and second instar nauplius. In fact, it has been questioned whether it is an excretory gland because

(1) the mesenchymal cells seen in the anlage may not form a schizocoel by the third-instar and (2) evidence for an excretory duct opening to the outside is lacking.

In summary, it appears that the prenaupliar embryo and first-instar nauplius lack the typical excretory glands. The second-instar nauplius may have a functional antennal gland, possibly acting as a renal organ.

B. ALIMENTARY TRACT

In the adult Croghan (1958d) has shown the midgut to be the site for replacement water taken up by active transport of NaCl. The replacement water for osmotic water loss comes from external fluids ingested at the mouth that percolate through the alimentary tract. In larval shrimp, Hootman and Conte (1974) have shown that the alimentary tract is not developmentally complete prior to the first larval molt. An important functional consequence of this finding is that, in order for the midgut to be capable of absorbing ions and water from ingested fluids, there must be some mechanism to allow percolation of fresh external fluids to replenish the absorbate taken from the midgut fluids. This would easily occur in the second-instar nauplius where an open passage exists between foregut and hindgut that allows for continuous flow of external fluids. In the first-instar nauplius, the foregut does communicate with the midgut, but there is a yolk plug blocking fluid movement into the hindgut. Fluid entrapped in this midgut cul de sac would require some kind of lavage (possibly provided by the movement of the labrum) to circulate fresh external fluids. The hindgut is open to the outside during the first-instar nauplius, but the epithelium does not contain the fine structural specializations that suggest an absorptive role (Hootman and Conte, 1974).

In summary, it appears that midgut epithelium is functional by second instar nauplius and may be the site for active uptake of NaCl and water.

C. CEPHALOTHORACIC ORGAN: "SECRETORY-TYPE GLAND"

Early studies (Leydig, 1851; Claus, 1876; Spannenberg, 1875; Zograf, 1905) of larval crustacean morphology reported on a cell complex that was very prominent and constantly found on the dorsal side of the cephalothorax region of the first-instar nauplius (refer to Fig. 3). Despite its being a rather conspicuous

---

FIG. 4.   (A) Scanning electron micrograph of late prenauplius with hatching membrane covering the nonfunctional salt gland. Shell of cyst (C), salt gland (Sg) ($\times$120); (B) scanning electron micrograph of late prenauplius with hatching membrane removed to reveal surface of the functional gland. Abdominal structure (Ab); second antenna (An); hatching membrane (HM); salt gland (Sg) ($\times$120). Taken from Conte *et al.* (1977).

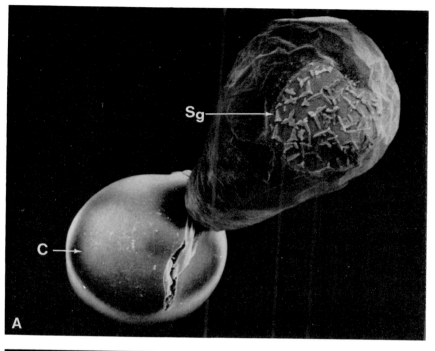

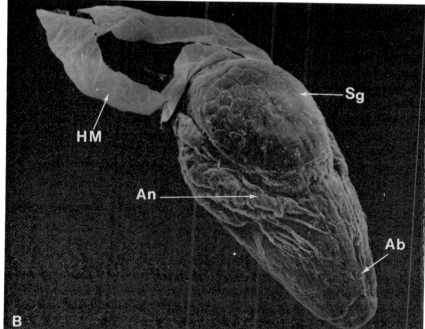

structure, most of the investigators were unable to establish a functional role for the cellular complex. Obviously they believed it had physiological significance, but they gave the organ various indifferent names, such as neck shield, neck organ, dorsal organ, median frontal organ, etc. Several possible functions have, however, been suggested, including (1) adherence to external objects (Zenker, 1851), (2) respiration (Dejdar, 1930), (3) support for the antennal and mandibular muscles (Weisz, 1947), and (4) salt extrusion (Croghan, 1958b; Conte *et al.*, 1972).

1. *External Morphology*

The cephalothoracic organ is prominent in the emerging prenaupliar embryo prior to hatching (Fig. 4A and B). During larval development, the epithelial layer is transformed from a flat, caplike structure into a hemispherical dome that has deep channels coursing through the epithelium. A thin cuticular film covers the epithelial cap, but a distinct ruffled edge serves as a boundary that identifies the limits of the organ (Hootman *et al.*, 1972). The organ measures between 150 and 200 μm across the diameter of the circular base and is composed of approximately 60 to 75 cells. These cells form a single layer of cuboidal epithelium and measure from 25 to 30 μm from the apical (cuticular) to the basal surface (hemocoelic).

2. *Histological and Histochemical Features*

Thick sections examined by light microscopy (Figs. 12A and 13A) show the gland arranged as a cellular complex with extensive interdigitations between adjacent cells. Each cell is characterized by a large nucleus with several prominent nucleoli, large numbers of yolk platelets, and mitochondria surrounded by a diffuse, membranous cytoplasm. Ultrastructural details and organellar characteristics are given in Section V.

For many years salt-secreting cells, especially those involved in chloride extrusion, have been qualitatively detectable at the light microscope level after appropriate histochemical treatment with fixatives containing silver ions for precipitation of chloride. When examined by light or scanning electron microscopy, the silver chloride (AgCl) precipitates appear as white granules. In the case of the first-instar nauplius, the AgCl granules are specifically located over the cephalothoracic organ (Fig. 5A and B). Prenaupliar embryos also show similar histochemical precipitation of chloride, as seen in Fig. 6A, except in this case the white granules of silver chloride have been replaced by deposits of free silver grains through the use of a reducing agent. To verify that the white or dark granular deposits are silver and/or chloride atoms, election diffraction and energy-dispersive analysis of the X rays emitted from the region overlying the organ have been made for both the prenaupliar embryo and first-instar nauplius (Fig. 6B). Similar findings have been reported for insect chloride cells and chloride epithelia found in aquatic insects (Komnick, 1977).

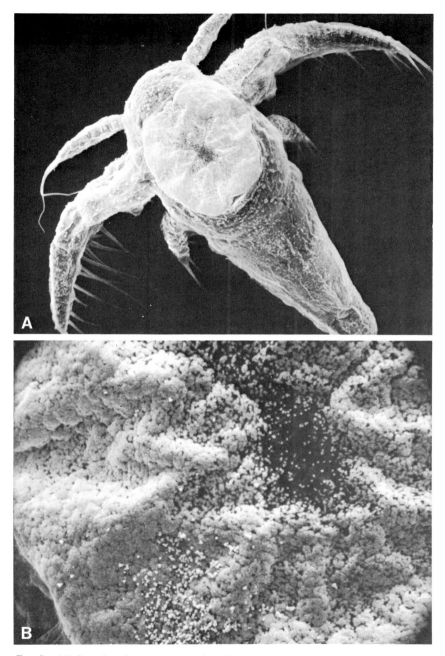

Fig. 5. (A) Scanning electron micrograph of first-instar nauplius with silver chloride granules lying on surface of salt gland (×135); (B) enlargement of surface of salt gland to show size of granules (×750).

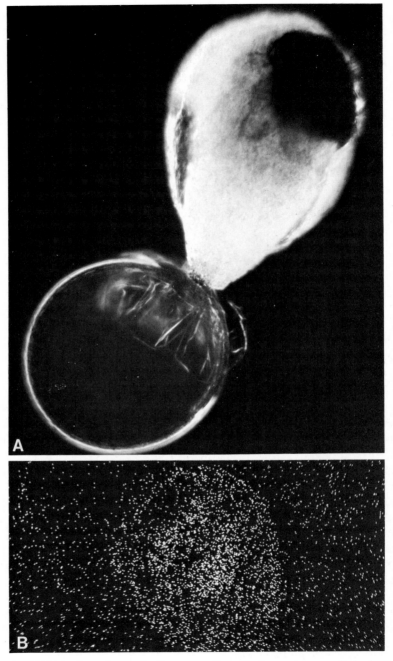

FIG. 6. (A) Light photomicrograph of late prenauplius with silver grains over salt gland ($\times 175$); (B) scanning electron micrograph of X-ray distribution image from silver atoms lying atop of salt gland from late prenauplius ($\times 175$).

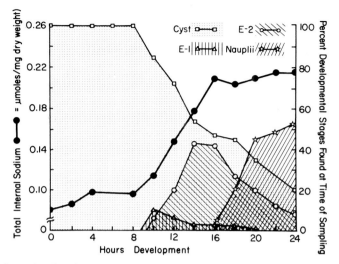

Fig. 7. Internal sodium levels of embryonic tissues during excystment of dormant eggs into free-swimming nauplii. Taken from Conte *et al.* (1977).

3. *Ionic Regulation and Chronology of Development*

The discovery of a "functional" salt gland in the prenaupliar stage embryo suggested that ionic regulation takes place very early in embryonic development. Prior experiments on hydration of dormant shrimp embryos (gastrula stage) showed that a passive mechanism for water balance existed. The passive osmotic mechanism utilized the manufacture of glycerol from trehalose to form hyperosmotic concentrations of free glycerol in tissue fluids that would allow bulk water to diffuse into the embryo. (Clegg, 1962, 1964). We were interested in finding whether sodium regulation ever took place in the embryo and, if it did, when it was occurring in the developmental chronology. Total body sodium (ECF) concentration was determined hourly throughout each larval stage (Conte *et al.*, 1977). The results shown in Fig. 7 indicate that sodium concentrations in ECF are low and constant in the dormant gastrula, but increase dramatically when the embryo breaks the chitinous shell and becomes directly bathed by the external medium. It can be seen that sodium concentrations reach a plateau and stabilize at the time of hatching and subsequently appear to be regulated throughout naupliar development. Therefore, the important transitional stage in development between the passive glycerol-mediated transport of water to the sodium-mediated transport processes lies somewhere in the prenaupliar stage.

4. *Inhibitors of Ionic Regulation*

Since the first-instar nauplius demonstrated rather remarkable sodium regulatory capabilities, we were interested in finding out if additional ion transport processes existed. Through the select use of metabolic inhibitors that blocked specific ion movements, toxicity studies would reveal the nature of these ionic transport processes.

a. *Ouabain: Inhibiting Sodium–Potassium Coupled Transport.* It was found that when ouabain, a specific inhibitor of Na^+, K^+-activated ATPase, was placed in the external saline medium, the nauplii demonstrated a high salt-dependent mortality (Fig. 8), a rapid rise in sodium concentration in the hemolymph (Table VI), and a fall in transepithelial potential across the epithelial layer of the cephalothoracic organ (Fig. 10). These findings suggested that the naupliar sodium regulation of the ECF is dependent upon the cationic transport enzyme.

b. *Acetazolamide: Inhibiting Chloride–Bicarbonate Coupled Transport.* Since chloride is an important ionic species involved in the outward flow of NaCl from salt extruding epithelia, we made the assumption that chloride movement would involve an active transport process, possibly enzymatically mediated. If

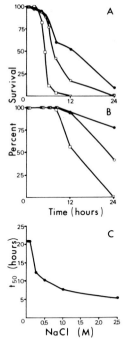

FIG. 8. Survival of nauplii in media of varying salinities containing ouabain: (A) 10^{-3} M ouabain; (B) 10^{-4} M ouabain; (C) time required for 50% mortality of nauplii (t_{50}) in media of varying salinities containing 10^{-3} M ouabain. Numbers of nauplii ranged from 100 to 300 for each time course. Controls containing no inhibitor showed greater than 95% survival in all cases. Media contained ●——●, 0.05 M NaCl; ○——○, 0.5 M NaCl; □——□, 2.5 M NaCl. Taken from Ewing *et al.* (1972).

TABLE I

Naupliar Lethality (LT_{50}) after Exposure to Dissolved Sulfonamide in NaCl-Fortified Seawater[a]

Sulfonamide (K_c[b] = dissociation constant in mol/liter)	Dosage (M)	Expts.[c] (n)	LT_{50} for salinity concentration (M, NaCl)				
			0.25	0.50	1.0	1.5	2.5
Sulfanilamide (K_c = 2.8 × 10^{-6})	10^{-3}	4	>36	>36	>36	>36	>36
Cyclothiazide (K_c = 6.0 × 10^{-7})	10^{-3}	4	>36	>36	>36	34±2	32.2±2
Dichlorphenamide (K_c = 2.5 × 10^{-8})	10^{-3}	3	>36	>36	33±3	30±2	29±3
Acetazolamide (K_c = 8.0 × 10^{-9})	10^{-3}	4	>36	>36	32±1	27±2	22±3
	10^{-5}	3	>36	>36	>36	30±2	
	10^{-6}	3	>36	>36	>36	>36	32±2
N-t-Butylacetazolamide[d] (K_c = 1, nonbinding)	10^{-3}	4	>36	>36	31±1	25±1	22±2
	10^{-4}	4	>36	>36	33±2	28±1	25±2
Ethoxzolamide (K_c = 1 × 10^{-9})	10^{-3}	4	>36	>36	31±2	24±1	20±2
	10^{-4}	3	>36	>36	31±2	28±2	21±1
	10^{-6}	3	>36	>36	>36	33±2	30±2

[a]LT_{50}, Length of time in hours to reach 50% mortality.

[b]K_c, Dissociation constant of sulfonamide complex with Zn^{2+} under equilibrium dialysis conditions (Coleman, 1975).

[c]Number of experiments utilizing 400–800 animals per exposure.

[d]N-t-Butylacetazolamide given as a gift from T. Muther, University of Florida.

the enzyme-coupled transport process were similar to the carbonic anhydrase (CA) mediated bicarbonate–chloride exchange mechanism as proposed by Maetz (1971), then an investigation into naupliar acetazolamide toxicity would be very useful because the mode of action of this sulfonamide drug is to bind zinc, and CA is a zinc-containing enzyme. A systematic study of naupliar toxicity for a variety of sulfonamides was initiated using a range of dosages (10^{-3} to $10^{-6} M$) and the results are depicted in Table I. This information led us to conclude that CA was present in the embryo. Unfortunately, our initial efforts to isolate and characterize the brine shrimp CA using conventional analytical techniques, such as spectrophotometric, manometric, and titrimetric, which have been so successful in CA assays for other crustacean tissues (Henry and Cameron, 1982; Burnett et al., 1981) proved to be unsuccessful. Similarly, when other investigations of vertebrate ion transporting tissues (i.e., turtle bladder, toad cornea, and bladder) found that their original CA enzyme assay produce negative results (Urakabe et al., 1976; Kitahara et al., 1967; Maren, 1967) they had to resort to

TABLE II
COMPARISON BETWEEN NAUPLIAR CARBONIC ANHYDRASE (CA) TO TISSUES ORIGINALLY
REPORTED LACKING CA ENZYME

Tissue[a]	Carbonic anhydrase (enzyme unit[b]/mg protein)	Reference
Turtle bladder, mucosal cells	1.03 ± 0.16	Scott et al. (1970)
Turtle erythrocyte	0.90 ± 0.19	
Toad bladder, mucosal cells	2.56 ± 0.46	Scott and Skipski (1979)
Toad erythrocyte	5.23 ± 0.40	
Brine shrimp nauplius	0.25 ± 0.10	Silverman (personal communication)[c]

[a]Tissues reported lacking carbonic anhydrase cited in Maren (1967), Kitahara et al. (1967), and Urakabe et al. (1976).

[b]One enzyme unit is the amount of enzyme that doubles the hydration rate (Maren, 1967).

[c]^{18}O-labeled bicarbonate exchange method using the low pH range (Silverman and Tu, 1976).

special techniques. These tissues now demonstrate the presence of low levels of CA activity, as shown in Table II (Scott et al., 1970; Scott and Skipski, 1979; Maren and Sanyal, 1983). Therefore, we made an extra effort and reexamined naupliar tissue using the ultraanalytical ^{18}O-exchange method (courtesy D. Silverman, University of Florida) for detecting miniscule amounts of CA, and found CA to be present at levels considerably lower than the CA activity present in epithelial cells reported in Table II. Lastly, as Maren (1967) suggested, in order to determine whether the loss of CA activity was responsible for the brine shrimp toxicity, it was deemed advisable to use a compound that was closely allied to one of the more potent sulfonamides and see whether the derivative exhibited similar salt-dependent toxicity as the parent compound. We chose to use N-t-butylacetazolamide because it lacks the ability to interact with CA due to the replacement of the proton on the amide group with an alkyl group (Maren, 1956) and found the LT_{50} values for N-t-butylacetazolamide to be nearly equal to acetazolamide (Table I). At the present time we feel that a CA-mediated HCO_3-Cl exchange does not exist due to the paucity of CA in naupliar tissue. The salt-dependent naupliar toxicity to N-t-butylsulfonamide is both reproducible and dose dependent indicating that we are not dealing with an artifact but the mechanism of action of the sulfonamides may be related more to the bioenergetics of ion transport rather than the blocking of CA. I shall make additional comments on this aspect of the problem when I discuss the bioenergetics of salt secretion in Section VI.

In summary, the developmental morphology together with the histochemical, physiological, and pharmacological evidence indirectly support the functional role of the cephalothoracic organ as being a "salt secretory" gland. The direct evidence establishing the physicochemical characteristics of the secretory fluid coming from the gland had to await the development of micropuncture techniques using a single nauplius. With these methods, it would be possible to (1) provide a microanalysis of the chemical composition of the naupliar hemolymph and (2) establish net movement of salts and water across the glandular epithelium by measurement of ionic fluxes and electrical gradients existing between hemocoelic compartment and environment.

III. Chemical Properties of Larval Extracellular Fluids

If we assume that the cephalothoracic organ acts as an extrarenal gland then it together with the antennal gland and the incompletely formed midgut can provide the growing nauplius with the excretory mechanisms necessary for osmotic and ionic regulation of extracellular fluids. Russler and Mangos (1978), in collaboration with my laboratory, successfully designed and manufactured a glass capillary trap for holding individual naupliar shrimp. The naupliar trap permitted micromanipulator controlled puncturing pipets to enter the anterior portion of the cephalothoracic hemocoel and remove nanoliter quantities of hemolymph from the cavity into the tip of the collection pipet. Upon removal the tip was placed under oil and stored until chemical and radiochemical analysis of the hemolymph could be performed. Details of the methods used in the chemical and radiochemical microanalyses has been published (Russler and Mangos, 1978). Similarly, we modified the naupliar trap to accommodate a triple-barreled inner glass micropipet that contained a central sensing electrode for the measurement of an open-circuit transepithelial voltage between hemocoel and external medium. Voltage-clamp techniques for measurement of short-circuit current across the epithelial boundary were unsuccessful *in situ* due to the (1) limited surface area, (2) small volume of the hemocoel, and (3) inherent elasticity of the cuticle covering the salt gland that caused the punctured nauplius to detach from sensing electrode when the shunting electrode was applied to the cuticular surface. Details of the transepithelial voltage techniques are given in Section IV.

A. Osmotic Concentration of the Hemolymph

Table III contains the hemolymph osmolarities taken from naupliar brine shrimp acclimated to various external media. The media used in these experiments were artificial sea water solutions which were fortified with additional amounts of NaCl. As can be seen from the data, naupliar osmoregulation is easily achieved throughout the wide range of salinities. Hemolymph osmotic concentrations are much lower than the external media, indicating that hypo-osmotic regulation is being maintained within the cephalothoracic coelomic compartment.

TABLE III

Osmotic Concentration of Naupliar Hemolymph from Brine
Shrimp Living in NaCl-Fortified Seawater Media[a]

Saline media (mosmol/liter)	Hemolymph (mosmol/liter)	Reference
155	118	Russler and Mangos (1978)
729	149	
932	161	
1510	197	
7000	194	
9114	224	
1056	413	Croghan (1958a)
932	160	Conte (1980)
2112	178	
3158	183	
4200	190	

[a]Values are means with 1 SD equal to ± 10%. Number of animals micropunctured per saline media ranged from 4 to 6. Nauplii acclimated in different saline media from 2 to 4 hours prior to micropuncture.

B. Ionic Comosition of the Hemolymph

Comparison of the major ions found in the hemolymph indicates that the nauplius has tremendous ability to maintain stable hypoionic concentrations of sodium, potassium, and chloride (Table IV). These ions are the major constitu-

TABLE IV

Comparison of Naupliar Ion Concentrations in Hemolymph from Brine Shrimp Living
in NaCl-Fortified Seawater Media[a]

Sodium concentration		Potassium concentration		Chloride concentration	
Hemolymph (mM)	SW bath (mM)	Hemolymph (mM)	SW bath (mM)	Hemolymph (mM)	SW bath (mM)
93	170	7	12	97	179
97	250	7	12	96	240
102	502	8	12	98	551
86	500	8	12	83	480
116	1000	8	13	87	989
110	1500	8	12	89	1486
93	2000	9	12	90	1980
109	2500	8	13	101	2487
115	4900	7	12	110	4900

[a]Values are means with 1 SD equal to ± 10%. Number of animals micropunctured for each saline bath ranged from 4 to 6. Nauplii were acclimated from 2 to 4 hours in each saline media prior to micropuncture. Data taken from Russler and Mangos (1978) and Conte (1980).

TABLE V

EFFECTS OF OUABAIN (10^{-4} M) ON SODIUM AND
POTASSIUM CONCENTRATION OF THE HEMOLYMPH OF
NAUPLIAR BRINE SHRIMP[a]

Time (hours)	Hemolymph	
	Na (mEq/liter)	K (mEq/liter)
0	107	3
1	147	7
3	158	11
6	182	13
12	321	14
24	434	14

[a]External medium contained (mEq/liter): Na, 459; K, 16; Cl, 502; Mg, 48; Ca, 10; SO$_4$, 31. Each value represents the mean of 7 measurements from individual nauplii. Maximum standard error for Na was 20 mEq/liter and K was 2 mEq/liter, respectively (courtesy of Prof. Mangos).

ents that account for the bulk of the osmotic properties of the hemolymph. Therefore, in order for the nauplius to maintain chemical constancy of the ECF found in the hemocoelic cavity, the glandular epithelium must possess cellular mechanisms that transport net amounts of NaCl from the cavity to the external environment (Table V).

IV. Transport Properties of Larval Salt Gland Epithelium

Measurement of transepithelial potential differences occurring across the glandular epithelium required modification of the naupliar trap. I shall describe in detail how the data were obtained: (1) *in situ* measurements of the open-circuit transepithelial potential differences (TEP) across the intact gland and (2) *in vitro* measurements of TEP across the glandular epithelium of isolated salt glands.

A. TRANSEPITHELIAL POTENTIAL DIFFERENCES: GLAND *in Situ*

1. *Micropuncture of Nauplius*

Several nauplii were transferred from the holding beaker containing either artificial hemolymph or appropriate artificial saline (Table VI) to a chamber on the stage of a stereomicroscope. A cone-shaped glass naupliar trap was fashioned from 3-mm glass tubing pulled on a micropipet puller and shaped with a micro-forge (Fig. 9A). The outer diameter of the large end of the core was 250–300 μm. The other end of the 3-mm glass tubing was fitted to a special metal adapter

TABLE VI

COMPOSITION OF ARTIFICIAL SALINES USED IN TEP MEASUREMENTS
ACROSS *in Situ* LARVAL SALT GLAND

Solution	Concentration (mM)										
	Na	K	Mg	Ca	cho[a]	Cl	HCO_3	methylS[a]	pH	NH_4	SO_4
Artificial hemolymph[b]	170	8	2	2	—	150	30	—	8.4	—	—
0.25 *M* sodium chloride	250	8	2	2	—	230	30	—	8.4	—	—
0.50 *M* sodium chloride	500	8	2	2	—	480	30	—	8.4	—	—
1.00 *M* sodium chloride	1000	8	2	2	—	980	30	—	8.4	—	—
1.50 *M* sodium chloride	1500	8	2	2	—	1480	30	—	8.4	—	—
2.0 *M* sodium chloride	2000	8	2	2	—	1980	30	—	8.4	—	—
2.5 *M* sodium chloride	2500	8	2	2	—	2480	30	—	8.3	—	—
0.48 *M* choline chloride	—	8	2	2	478	480	30	—	8.6	22	—
0.24 *M* choline chloride	239	8	2	2	239	480	30	—	8.3	22	—
0.50 *M* sodium methylS	500	8	2	2	—	—	30	478	8.3	—	2
0.24 *M* sodium methylS	478	8	2	2	—	239	30	239	8.3	—	2

[a]methylS, methylsulfate; cho, choline.
[b]Modified from Smith (1969a).

which accommodated a triple-barreled inner glass micropipet. The adapter was attached to a glass syringe that was held by a micromanipulator. The inner glass micropipet was composed of a central sensing electrode (or perfusion pipet) to which an exchange pipet and a pressure pipet were attached (Fig. 9D). This single multibarreled pipet served as both a capture and micropuncturing device. The pressure pipet permitted flushing of the sensing electrode with artificial hemolymph to ensure the tip was open and free from debris. This flushing action, which allowed excess fluid to bypass the sensing electrode and go out the exchange pipet, also permitted easy capture of a single nauplius. The fluid bypassing the sensing electrode created a high velocity stream effect inside the capture cone which caused any nauplius swimming near it to be swept inside and be captured (Fig. 9B). Once inside the cone, the nauplius was immobilized by the slight negative pressure created by the pressure pipet. A reversal of the fluid flow in the pressure pipet caused the restrained nauplius to be discharged from the capture cone. Many animals captured and released in this fashion continued to swim and function in a normal behavior pattern for at least 6 hours.

The outer pipet and inner micropipets had separate control systems which made it possible to move each independently of the other. During the capture process, the inner micropipet (sensing electrode) was retracted from the tip of the cone to prevent accidental impalement of the animal. After capture of the animal, it was a simple matter to lower the sensing electrode and to micropuncture the salt gland if it was oriented in a suitable position (easily established by observing the position of the orange pigmented anteriordorsal ocellus). After micropuncture, the coordinated movement of the inner sensing electrode and pressure pipet allowed the animal to be released from the capture cone while it was still tethered to the sensing electrode (Fig. 9B and D and 9C and E). Animals were maintained in this position with the locomotory antennae beating synchronously and rapidly for up to 4 hours without apparent electical leakage. The fluids in the bath chamber were rapidly changed by means of two synchronized 10-ml glass syringes which would remove and replace the bath fluid at the same rate, thereby maintaining constant bath volume (0.8 ml) and constant hydrostatic pressure.

2. Electrical Circuitry

The open-circuit transepithelial voltage was measured through the micropipet. Agar bridges (4% agar in artificial hemolymph solution) connected the ends of the sensing-perfusion pipet and the bath to separate saturated KCl solutions. The electrical circuit was closed with equivalent Calomel half-cells (Corning) in series with a voltmeter (Keithley 602) and a voltage reference source (W-P Instruments, model 101). Voltage asymmetries usually were less than 1–2 mV and were measured in artificial hemolymph solution. In the analysis of the data, corrections for liquid junction potentials were disregarded because of their small size in comparison to the magnitude of the transepithelial potentials.

3. Sensing Electrodes

Micropipets with a tip diameter of about 1–2 μm were pulled from 2-mm glass capillary tubing, with a microelectrode puller. Each tip was beveled to an angle of 30° on a glass polishing wheel to facilitate puncturing the cuticle. After being polished, the electrodes were filled with artificial hemolymph and placed inside the outer capture pipet held by the metal adapter.

4. Transepithelial Potential Difference in Artificial Hemolymph and Other Solutions

Individual naupliar preparations were judged to be acceptable when (1) the nauplius was demonstrating excellent locomotory activity while tethered on the microelectrode, (2) the salt gland generated a hemolymph-positive transepithelial potential greater than 10 mV with the nauplius bathed in artificial hemolymph, and (3) the potential did not change by more than 1 mV/minute during the recording time of 15–30 minutes. Likewise, new steady-state potentials following changes in the composition of the bath fluid were defined as potentials which did not change by more than 1 mV/minute. Figure 10 shows the transepithelial potential across the *in situ* larval salt gland epithelium as a function of the concentration of sodium in bathing solution. It can be seen that the TEP increases as the concentration of sodium in the bath is elevated. The measured TEPs as seen in Fig. 9F are lower than steady-state equilibrium potentials (E_{Na}), calcu-

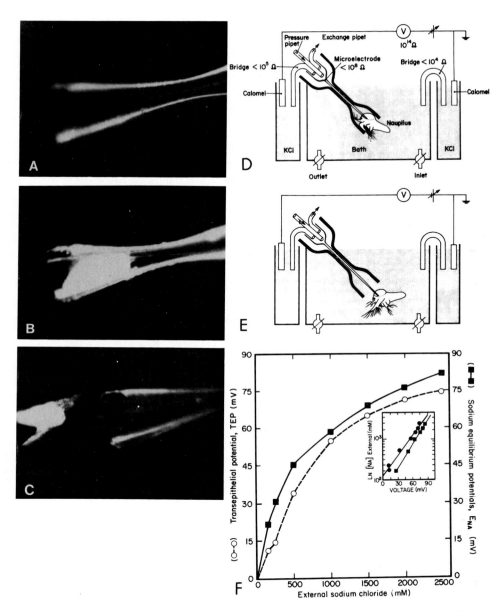

FIG. 9. Light micrograph of cone-shaped glass naupliar trap and schematic diagrams showing multibarrel micropuncture pipet, microelectrode, and electrical circuitry. (A) Capture cone of outer pipet surrounding the inner microelectrode [perfusion-sensing pipet ($\times 50$)]; (B) nauplius restrained in capture pipet ($\times 50$); (C) nauplius being measured while tethered to sensing microelectrode ($\times 50$); (D) schematic diagram showing nauplius during capture and holding situations in bath chamber; (E) schematic diagram showing nauplius tethered-swimming and measurement situations in bath chamber; (F) transepithelial potentials as a function of external salinity. ($\bigcirc$ measured potential (TEP); ($\blacksquare$) equilibrium potential (E_{Na}). Values are means of measurements taken from five or more individuals. Standard error for each mean is less than $\pm$ 2.0 mV.

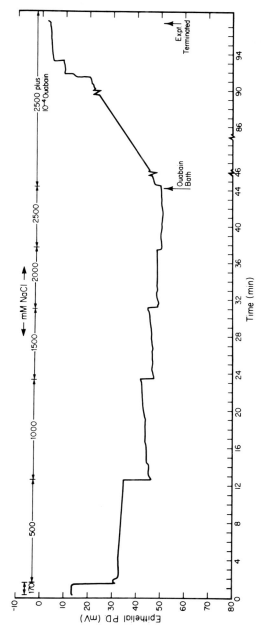

FIG. 10. Inhibitory effects of ouabain on electrochemical potential across salt gland epithelium as evidenced in recorder trace of TEP.

67

TABLE VII

MEASUREMENT OF TRANSEPITHELIAL POTENTIAL DIFFERENCE ACROSS
in Situ LARVAL SALT GLAND EPITHELIUM AS A FUNCTION OF
EXTERNAL SODIUM CONCENTRATION

Sodium concentration[a] (mM)		Potential difference[b] (mV)	
Hemolymph	External saline	E_{Na}	TEP[c]
97	170	14.0	10.9 (15)[d]
96	250	24.7	14.1 (19)
86	500	45.4	34.4 (7)
116	1000	55.7	55.3 (4)
110	1500	67.5	64.3 (4)
93	2000	79.2	71.3 (4)
109	2500	80.9	74.0 (4)

[a]Sodium values are taken from Table VI.

[b]Measured PD has hemolymph surface at positive (+) polarity. Maximum standard error for TEP was ±1.0 mV.

[c]TEP measurements were made in Prof. W. Dantzler's laboratory, University of Arizona; I am indebted to Dr. Klaus W. Beyenbach and Ms S. Bentley for technical assistance.

[d]Number of animals micropunctured.

lated from values of hemolymph sodium concentrations of nauplii in similar external concentrations (Table VII). One criticism directed toward measurements of TEP exhibiting small differences from the E_{Na} is the difficulty in providing evidence of a perfect seal around the internal electrode. If a leak had occurred around the microelectrode in those animals being bathed in high salt solutions, then TEP measurements should have been unstable. However, as shown in Fig. 10, the TEP was quite stable even with very high NaCl concentrations in the bath.

Measurement of TEP showed changes when the concentration or composition of the external bath was altered or when pharmacological agents were added. Replacement of sodium by choline caused a large decrease (>90%) in TEP as did the addition of 10^{-4} or 10^{-6} M ouabain in the presence of sodium (Fig. 10). Substitution of methyl sulfate for chloride did not produce a change in the TEP. However, a amiloride (10^{-4} or 10^{-5} M) produced definitive decreases in TEP which were dependent upon salt concentration in the external bathing fluid.

B. TEP DIFFERENCES AND IONIC FLUXES: ISOLATED SALT GLAND

1. *Single Larval Salt Gland*

Details of a successful protocol for the batch removal of intact larval salt glands from living nauplii together with the specifications for maintaining cellular integrity and functionality have been reported (Lowy, 1984; Lowy and Conte, 1984). Briefly, it involves the incubation of nauplii in a phosphate-enriched balanced salt solution (SGBM), a so-called "Crustacean Ringers" solution, for 8 hours at an elevated but nonlethal temperature (37°C). Following incubation, the phosphate-treated nauplii are vortexed in a test tube containing antibiotic–SGBM and the shearing forces generated by vortexing causes fracturing of the cuticular boundary surrounding the gland. The crude salt gland mixture is separated from cellular and tissue debris by passing the antibiotic–SGBM–salt gland mixture through a series of fine nylon mesh screens to remove extraneous material. Concentration of the diluted SGBM–salt gland mixture is achieved by passing the solution over a glass-beaded column that causes other epithelial cells to adhere but allows passage of free glands. The protocol results in an enriched isolate yielding 5000 glands per 28 g of nauplii. Morphological integrity of the glands was ascertained by electron and light micrography (Fig. 11A, B, and C). Cellular functionality was demonstrated through the use of a basic battery of assays, such as (1) the vital dyes (trypan blue, eosin, nigrosin) being excluded by the glandular epithelial cells for at least 24 hours even when glands were maintained in simple salt solutions; (2) oxygen consumption being maintained at a steady rate of approximately 22.0 nM O_2/minute/mg protein for several hours and which could be predictably altered by compounds known to effect oxidative phosphorylation (CN^-, DNP, etc); (3) having the presence of large quantities of cationic transport enzyme, Na^+,K^+-ATPase, with a high specific activity (9.1 mM P_i/hour/mg protein).

2. Media Composition for TEP and Sodium Flux Measurement on Isolated Glands

The composition of the crustacean basal saline solution used to measure TEP and ^{22}Na fluxes on isolated salt glands consisted of 85 mM NaCl, 5 mM KCl, 5 mM NaHCO$_3$, 5 mM MgCl$_2$, 1 mM CaCl$_2$, and 10 mM sodium phosphate buffer, pH 7.6. Isotopic stock solution, as ^{22}NaCl, was added to the saline solution bathing one side of the epithelium after equivalent volume and concentration was removed and used as a background sample. After equilibration time of 30 to 60 minutes, 10-μl samples were taken from the cold side every 5 minutes and 2-μl samples from the radioactive hot side. Assays were made by use of a scintillation cocktail and counted in a liquid scintillation counter.

3. Micromanipulation of Isolated Gland for Open- and Short-Circuited TEP

Measurement of short-circuited TEPs, current, and conductance required the use of a double-barrel theta glass perfusion micropipet with a tip polished to approximately 100 μm diameter (Fig. 12). Each gland was positioned into a "capture position" using a vacuum-perfusion microelectrode held by a Prior micromanipulator under a Zeiss SR-1 dissecting scope at 50× magnification and attached by the cuticular side with a "vacuum-sealant system." Epithelial cur-

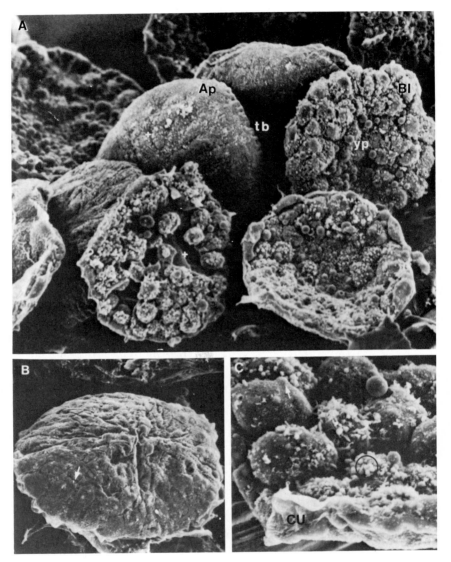

FIG. 11.   Scanning electron micrographs of the apical (Ap) and basolateral (Bl) boundaries of the isolated salt gland with details of cell surface. (A) The flexibility of the cuticle (CU) can be deduced from the folds (+) seen from both the apical and basolateral surface. The transition band (tb) appears as a crenulated edge. Yolk platelets (yp) appear as smooth spheres (×450); (B) apical view showing cell borders clearly (arrows (×540); (C) higher magnification showing cuticle edge (CU) and the approximate spherical shape of individual cell bodies. Bulges in the surface (small arrow) probably correspond to internal yolk platelets. Numerous small projections (circle) may correspond to portions of the tubular labyrinth (×880). Taken from Lowy (1984).

rents and TEPs were measured when the cuticular-tip seal resistance was found to be in the range of 10–100 MΩ. A whole cell/patch clamp instrument (WPI– Model FPC-7) was used for these experiments which permits simultaneous measurements of epithelial current (pA) and conductance (10 pS range) at specified potentials. Initial experiments utilizing a single-barrel nonperfused micropipet to compare open-circuit TEP and sodium flux measurements between the isolated and intact gland gave the following results. As can be seen in Table VIII, the average volume obtained for the open-circuited TEP is 31.3 ± 1.2 mV with the basolateral side positive in polarity and compares favorably with the value of 34.2 ± 2.6 for the intact animal. However, the ^{22}Na flux measurements indicates that the net flux from the isolated gland is slightly lower (77%) than that of the intact gland. Isotopic chloride influxes and effluxes have not been performed. It must be cautioned that these measurements are considered as preliminary since the experimental "vacuum-sealant system" is not completely perfect.

4. Active Sodium and Chloride Transport

Can passive electrochemical forces account for sodium and chloride movement across the larval salt gland epithelium? If the movement of sodium across the epithelium of the larval salt gland occurs entirely by passive diffusion, one would expect a ratio of the unidirectional sodium permeabilities to be unity. It is important, therefore, to compare the ratio of the outward to inward permeability constants for sodium as derived from the Goldman equation (Goldman, 1943) for both the isolated and intact gland. For the intact gland, the calculated ratio of the sodium permeabilities is

$$\frac{P_o^{Na}}{P_i^{Na}} = \frac{1.04 \times 10^{-5} \text{ cm sec}^{-1}}{4.46 \times 10^{-6} \text{ cm sec}^{-1}} = 2.53$$

while the ratio for the isolated gland is

$$\frac{P_o^{Na}}{P_i^{Na}} = \frac{6.30 \times 10^{-6} \text{ cm sec}^{-1}}{3.99 \times 10^{-6} \text{ cm sec}} = 1.58$$

The sodium permeability constants are obviously different from unity and suggest an active, outward-oriented transport of sodium. This finding is compatible with electrical measurement across the glandular epithelium which show TEPs smaller than the sodium diffusion potential (E_{Na}). Unfortunately, the short-circuited current and conductance have not been measured on the isolated gland; therefore, comparison of the sodium partial conductance to the total measured conductance cannot be made.

Isotopic chloride influxes and effluxes have not been performed on isolated larval salt glands, but one must assume a flux ratio near 1 under steady-state conditions if chloride movement is passive. However, using $C_o^{Cl} = 98$ mM and

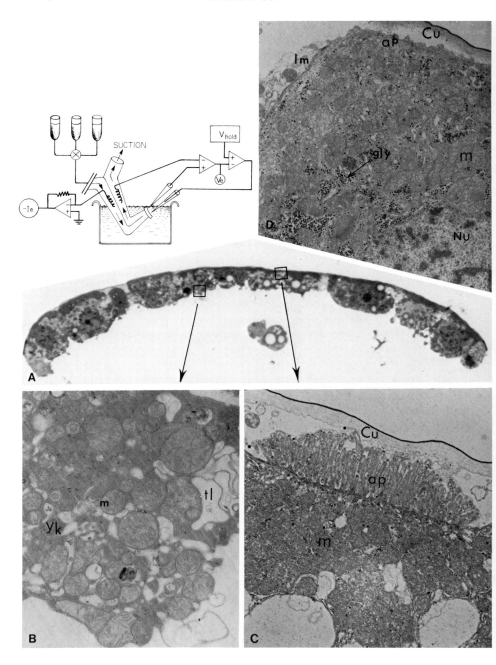

TABLE VIII

COMPARISON OF SODIUM FLUXES, POTENTIAL DIFFERENCES, AND APPARENT SODIUM DIFFUSIONAL
PERMEABILITY ACROSS *in Situ* AND *in Vitro* LARVAL SALT GLAND EPITHELIUM

Expt. conditions[a]	Unidirectional sodium fluxes (pmol/cm²/sec)		Transepithelial potential[b] (mV)		Sodium diffusional permeability[c] (cm/sec)	
	J_i^{Na}	J_o^{Na}	E_{calc}	E_{obs}	P_o^{Na}	P_i^{Na}
Salt gland, *in situ* cephalothorax region C_o = 500 mM [Na] C_i = 86 mM [Na]	1060 (7)[d]	1621 (7)	+45.2	+34.4	1.04×10^{-5}	4.46×10^{-6}
Salt gland, *in vitro* with hemolymph surface bathed by crustacean Ringer's C_i = 120 mM [Na] C_o = 500 mM [Na]	950 (3)	1380 (3)	+45.2	+31.3	6.30×10^{-6}	3.99×10^{-6}

[a]*In situ* salt gland surface area calculated to be 6.28×10^{-4} cm². *In vitro* salt gland surface area beneath micropipet calculated to be 1.57×10^{-4} cm².

[b]Measured PD has fluids bathing basolateral surfaces at positive (+) polarity.

[c]Calculations of apparent sodium diffusional permeability taken from the Goldman equation (Goldman, 1943).

[d]Number of animals micropunctured or glands successfully mounted.

C_o^{Cl} = 551 mM taken from Table IV and a measured TEP of +34.4 mV, we calculate a flux ratio of J_{out}/J_{in} = 40.1 (Ussing, 1949). In addition, the measured TEP is markedly different from the electrochemical equilibrium potential for chloride which is −53.6 mV. Therefore, simple passive diffusion for chloride movement is unlikely and outward active transport is the most probable mechanism for chloride distribution.

In summary, most of the direct experimental evidence shows (1) the naupliar ECF being hypoosmotic and hypoionic to the various external media, (2) the electrochemical characteristics of the salt gland epithelial transport being depen-

FIG. 12. Schematic diagram of short-circuit apparatus and vacuum-sealant system for *in vitro* measurement of TEP across isolated salt gland. (A) Light micrographs of isolated salt gland epithelium together with the schematic of short-circuit apparatus (×100); (B) TEM of basolateral surface showing tubular labyrinth (tl), mitochondria (m), and yolk platelets (yk) (×8700); (C) TEM of apical surface showing extended tuft of tubular projections (ap) lying beneath endocuticle (Cu) and arising from mitochondria-rich (m) cytoplasmic zone of the chloride cell (×8700). (D) TEM of lateral border (lm) showing the tubular projections (ap) beneath the cuticle (Cu) adjacent to a spot desmosome and a lengthy septate junction. Nu, nuclei; m, mitochondria; gly, glycogen fields (×8700).

dent upon active extrusion of chloride and a small active component for sodium, and (3) epithelial cells of this structure being interpreted as active chloride secretory cells.

V. Crustacean Chloride Cells

A. ULTRASTRUCTURAL FEATURES OF ADULT AND LARVAL CHLORIDE CELLS

In an earlier paper (Conte, 1980), I defined the crustacean chloride cell as pertaining only to the epithelial cell found in the extrarenal organs of the brine shrimp; namely, the larval salt gland found in the cephalothoracic region of the naupliar brine shrimp and the salt organ of the adult middle leg segments. Ultrastructural studies of both the adult and larval chloride cells have demonstrated a large amplification of the surface area of the plasma membrane, a hallmark of epithelial cells destined to transport electrolytes (Copeland, 1966, 1967; Kikuchi, 1972; Hootman and Conte, 1975). What appears to be unique regarding the crustacean chloride cell is that two distinct allomorphic patterns have evolved from the surface boundary of the plasma membrane. From the basolateral surface comes the discrete tubular labyrinth that is closely associated with the mitochondrial complexes (Figs. 13B and 14B) and forms by the invagination of the plasma membrane juxtaposed to the outer mitochondrial membrane. This mitochondrial–labyrinth association can be found in both adult and larval chloride cells. In contrast, the architecture of the plasma membrane at the apical surface appears to extend into large numbers of irregularly shaped projections forming tubular tufts that lie beneath the cuticular surface in a compressed fashion (Figs. 13 and 14A) or become clearly extended when viewed in an isolated gland (Fig. 12C). The tubular tufts do not appear to be associated with mitochondrial complexes. However, on the apex of each projection there are small granular complexes which may contain chitin-precursor substances used in the formation of new cuticle. The apical surface is separated from the basolateral surface by an extensive septate junction (Fig. 15). Additionally, the entire epithelial complex of the gland does not rest upon a basal lamina (Fig. 13A and B) but derives its mechanical support from the cuticular-apical tuft interlacing. Larval chloride cells have two unusual ultrastructural features not found in the

---

FIG. 13.   Ultrastructural features of the crustacean chloride cell. (A) Simple epithelium of the salt gland with arrows indicating tubular tufts (ap), nuclei of chloride cells (nu), hemocoeloic cavity beneath salt gland (H), and overlying the anterior digestive diverticulae (Di) ($\times$1200); (B) TEM of transverse section through the salt gland. Portions of two cells are visible. Note cytoplasmic stratification. Nuclei (nu) and glycogen fields (gly) with yolk platelets (yk) near the hemocoelic surface (he). A labyrinth of tubular plasma membrane occupies the central and basal zones (tl) ($\times$4800). Taken from Hootman and Conte (1975).

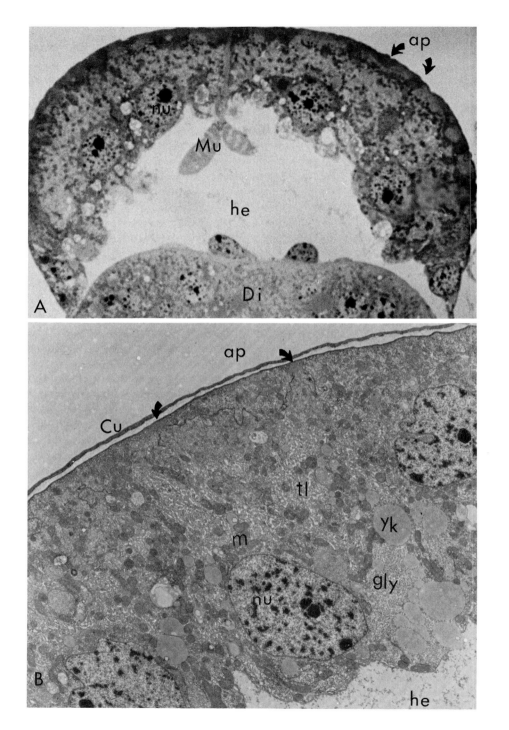

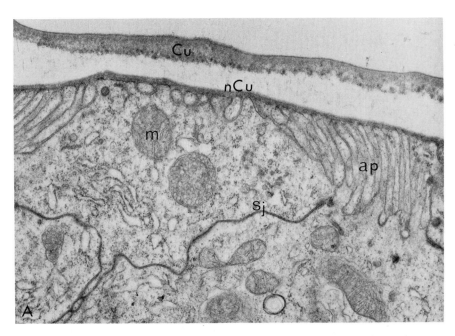

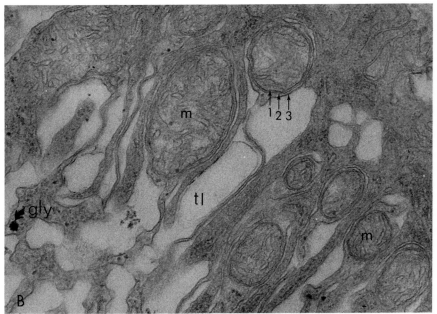

adult chloride cells. They are (1) the presence of large numbers of yolk platelets which contain a unique storage compound, a symmetrical pyrophosphate ester, P^1, P^4-diguanosine-5'-P^1,P^4-tetraphosphate (commonly referred to as diGDP or Gp_4G), and (2) large quantities of glycogen granules (Fig. 16A and B). These two features provide the cell with unique metabolic advantages for ATP production beyond that furnished by oxidative phosphorylation and will be elaborated upon in Section VI.

B. Plasma Membrane Microdomains in Larval Chloride Cells: Polarity of Na^+,K^+-ATPase

Molecular mechanisms which involve extrusion of sodium ion from epithelial cell cytoplasm directly into the external environment in exchange for environmental K^+ ion have required the placement of the transport membrane enzyme, Na^+,K^+-ATPase, on the apical surface of secretory cells (Kirschner, 1977, 1979). In vertebrate chloride cells, various cytochemical and immunocytochemical studies (see review by Ernst and Hootman, 1981) have provided evidence against an apical membrane location because most (if not all) of the Na^+,K^+-ATPase found in these cells is located on the basolateral infoldings of plasma membrane surface. In no case is there convincing evidence for any enzyme on the apical membrane. Given this intracellular location, the enzyme would not have direct access to environmental K^+ but only to the K^+ ions found in the ECF harbored in the infolding and/or tubular channels. Therefore, the enzyme would not be eliminating Na_{cyt}^+ from the cytoplasm but only returning Na_{cyt}^+ back into the blood (Na_{ECF}^+). It has been suggested that the net extrusion of Na_{ECF}^+ is passive and can be accomplished by electrical coupling of the Na_{ECF}^+ (presumably through paracellular or intracellular channels) with the active transport of chloride (Kirschner, 1979).

There has been little electrochemical evidence to support active transport of sodium in vertebrate chloride cells. However, if our measurements and interpretation of the electrochemical gradient existing across the larval salt gland epithelial complex are valid, indicating that a portion of the net extrusion of Na is active, then molecular mechanisms that would account for this portion of sodium movement might involve an entirely different placement of Na^+,K^+-ATPase on plasma membrane surfaces. Unfortunately, the subcellular localization of

---

Fig. 14. (A) Apical zone near a chloride cell boundary which can be identified by the long septate junction (Sj) and adjacent to numerous projections of tubular tuft (ap) beneath the new endocuticle (nCu). Mitochondria (m) are present in apical cytoplasmic zone ($\times 24{,}500$); (B) mitochondrial "pump" at the basal surface. Note the presence of the outer plasma membrane (3) of the tubular labyrinth (tl), juxtaposed to the outer (2) and inner (1) mitochondrial membranes. Note the presence of glycogen granule (gly) ($\times 49{,}350$).

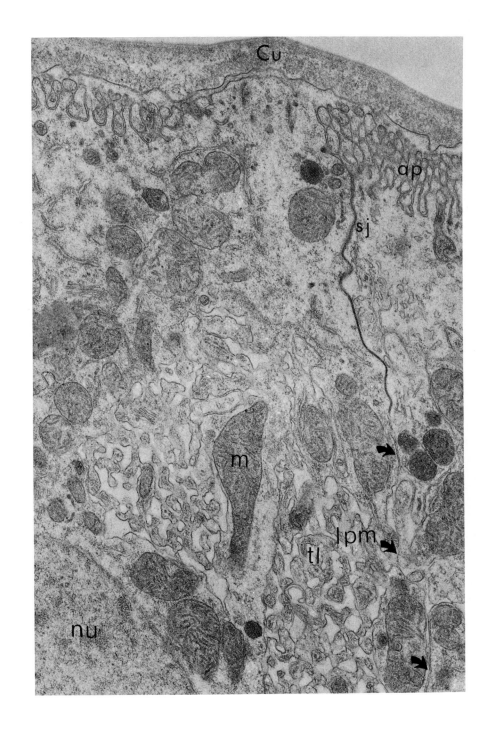

Na$^+$,K$^+$-ATPase had not been accomplished for the crustacean chloride cell until very recently (Lowy *et al.*, 1981).

1. *Brine Shrimp Na$^+$,K$^+$-ATPase: Two Epiforms of Na$^+$,K$^+$-ATPase*

As mentioned earlier (Section II,C,3) the NaCl-dependent ouabain toxicity alerted our laboratory to the involvement of the Na$^+$,K$^+$-ATPase in the ontogenetic development of naupliar ion regulation. Developmental studies on the biogenesis of naupliar membranes provided results which showed hemolymph sodium regulation to be dependent upon newly synthesized Na$^+$,K$^+$-ATPase (Conte *et al.*, 1977; Peterson *et al.*, 1978a,b). In addition, other membrane marker enzymes, such as 5" UMPase, glucose-6-phosphatase, and rotenone-insensitive NADH oxidase were heavily synthesized, indicating that extensive plasma membrane differentiation was occurring.

Therefore, I proposed a working hypothesis for studying the cytodifferentiation of the crustacean chloride cell. It would focus on the transcription and translation of the Na$^+$,K$^+$-ATPase genes with the attendant modification, transport, and insertion of the enzyme subunit polypeptides into newly formed plasma membrane. It was hoped that the salt gland system would give us some insight into the nature and growth of the tubular labyrinth and apical tufts from newly formed plasma membrane subunits. In order to conduct these kinds of experiments it became necessary to achieve two technical breakthroughs: (1) a protocol for the purification of the enzyme which allows the characterization of the molecular and kinetic properties of the Na$^+$,K$^+$-ATPase, and (2) a protocol for the separation of intact extrarenal organs such as the larval salt and midgut intestines that would provide primary organs for tissue culture.

Purification of the Na$^+$,K$^+$-ATPase from brine shrimp has been a formidable challenge but the most recent purification scheme (Peterson *et al.*, 1978b, 1982a,b) involves four steps.

1. A crude membrane fraction is first isolated from the homogenate by differential centrifugation.

2. The crude membranes are then fractionated by sucrose gradient centrifugation either in swinging bucket rotors for small-scale preparation (10–100 g tissue) or in zonal rotors for large-scale preparation (up to 1 kg tissue).

3. These gradient membranes are treated with a Lubrol WX cocktail and pelleted by centrifugation.

FIG. 15. Discrete septate junction (Sj) adjoins two chloride cells and lies above the central cytoplasmic zone. Arrows show lateral plasma membrane (lpm) separating the cells and invaginating to form the tubular labyrinth (tl) lying next to numerous mitochondria (m). Cuticle (Cu) locates the apical surface (×17,000).

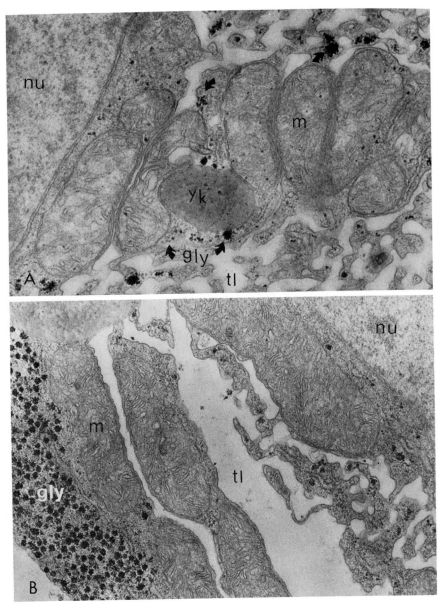

FIG. 16.   (A) Mitochondrial–yolk platelet complex showing the spatial nearness of yolk platelet (yk) to several mitochondria (m). Note the arrows depicting glycogen granules (gly) trapped between mitochondria, yolk, and tubular labyrinth (tl) membranes adjacent to the nucleus (nu) (×25,200); (B) mitochondrial–glycogen field complex showing granules of glycogen (gly) adjacent to mitochondria (m) and tubular labyrinth (tl) ×24,200).

4. The Lubrol pellet is treated with an SDS cocktail and fractionated by sucrose density gradient centrifugation in swinging buckets (small scale) or zonal rotors (large scale). The final particulate SDS-treated enzyme is isolated from the excess sucrose by centrifugation or ultrafiltration and centrifugation.

Results of a representative purification protocol are shown in Table IX. The purified enzyme when analyzed by SDS–PAGE yielded about eight bands. Three bands were found to enrich during the purification procedure, which included two closely spaced bands near $M_r = 100,000$ and one diffuse band near $M_r = 40,000$ (Fig. 17). Both of the larger molecular-weight bands were identified as α-subunits by phosphorylation with $[\gamma\text{-}^{32}P]ATP$ in the presence of Na^+ and Mg^{2+}, as shown in Fig. 18 (Peterson et al., 1978a; Peterson and Hokin, 1982). Identification of the β-subunit was confirmed by photoaffinity labeling with $[^3H]4'''\text{-}[2\text{-ethyldiazomaloynl}]$ digitoxin. The finding of two molecular forms of the phosphorylated subunit ($\alpha 1$ and $\alpha 2$) was quite startling in light of the fact that it had not been reported from other sources of purified Na^+,K^+-ATPases. Molecular properties of the brine shrimp Na^+,K^+-ATPase are shown in Table X and are similar to the reported values for other Na^+,K^+-ATPases. However, the α-subunit had always been assumed to contain no carbohydrate as determined by the lack of staining with periodic acid–Schiff reagent (PAS). But direct analysis of amino acid and carbohydrate composition of the purified brine shrimp Na^+,K^+-ATPase subunits clearly show that both the α- and β-subunits contain carbohydrate. Neither of the brine shrimp subunits contains sialic acid, whereas sialic acid is a constituent in both subunits from other sources (Peterson and Hokin, 1980; Churchill et al., 1979). Rates of ^{14}C-labeled amino acid incorporation into the α-subunit polypeptide demonstrated that the two forms are

TABLE IX

PURIFICATION OF Na^+,K^+-ATPase FROM BRINE SHRIMP NAUPLII[a]

Purification step	Total activity (μmol P_i/hour)	Total protein (mg)	Specific activity (μmol P_i/hour per mg protein)	Enzyme yield (%)	Relative purification (fold)
Homogernate	9720	3490	2.8	100	1.0
Crude membranes	4950	380	13.8	51	4.7
Gradient membranes	2480	51.6	48.1	26	17
Lubrol pellet	3580	27.5	130	37	47
SDS-treated enzyme	2370	4.4	534	24	191

[a]Taken from Peterson et al. (1982a).

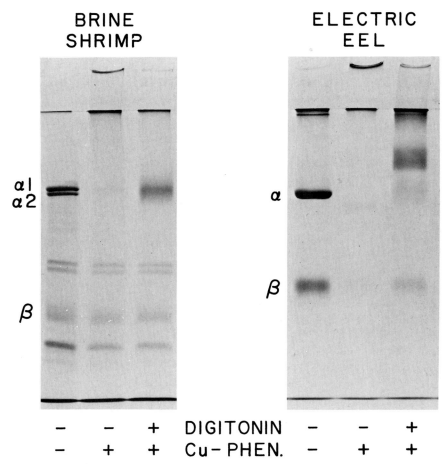

Fɪɢ. 17. Cupric phenanthroline (Cu-PHEN.) cross-linking of the NA^+,K^+-activated ATPase subunits from purified brine shrimp and electric eel enzyme. Each gel contains 50 μg total enzyme protein. Lanes 1 and 4 are control gels. Lanes 2 and 5 are Na^+,K^+-activated ATPase cross linked with cupric phenanthroline for 30 minutes at 37°C. Lanes 3 and 6 are Na^+,K^+-activated ATPase treated with 2.5 mg/ml of digitonin prior to cross-linking with cupric phenanthroline for 30 minutes at 37°C. Taken from Peterson *et al.* (1982a).

synthesized and degraded at similar rates throughout naupliar development. This finding, together with cell-free translational studies (Baxter *et al.*, 1983) in which the putative pre-α-subunit was found to consist of a single protein band intermediate in migration between mature α1 and α2, is inconsistent with any precursor–product relationship between α-subunits. Differences in α1 and α2 may lie in the polysaccharide component. Since the whole embryo was prepared as the naupliar homogenate for the isolation of crude plasma membranes, it is possible that the two molecular forms are present in several types of tissues or cells. A situation has recently been reported for rat brain Na^+,K^+-ATPase where two molecular forms of the α-subunit are also found (Sweadner, 1979). Each α-subunit has somewhat different primary polypeptide structure and differing substrate and cardiac glycoside binding sites. Localization and separation of α-subunits from various kinds of cultures of nervous tissue showed the α1 (α+) was derived from the plasma membrane of neuronal cells where the axolemma was stripped away from the myelin and α2 (α) was found primarily in non-neuronal cells (astrocytes). The interpretation of the enzyme isoforms being solely derived from these two different cell types was in conflict with the added finding that neuronal synaptosomal preparations contained α1- and α2-subunit forms. For example, a single myelinated neuron could have yielded both α1- and α2-subunits in which the presynaptic (α1) and postsynaptic (α2) membranous processes were derived from the same cell body.

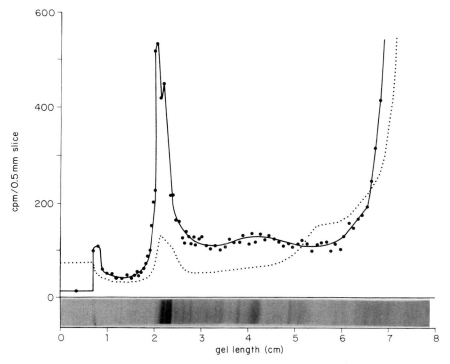

FIG. 18. Distribution of radioactivity in SDS polyacrylamide gels prepared from FT-membrane enzyme phosphorylated with [γ-^{32}P]ATP in the presence of Na (solid line) or K (broken line). A coomassie Brilliant Blue stained Na gel is shown below. Taken from Peterson *et al.* (1978a).

TABLE X

MOLECULAR PROPERTIES OF BRINE SHRIMP Na$^+$,K$^+$-ATPase[a]

Subunit large (α) (g/mol)	M_r small (β) (g/mol)	Subunit ratio (large/small) Mass	Subunit ratio (large/small) Molar	Specific activity (μmol P_i/hr/mr prot)	Ligand binding data [γ-32P]ATP MA (min^{-1})	[γ-32P]ATP M_r (g×10^{-3}/mol)	[3H]Ouabain MA (min^{-1})	[3H]Ouabain M_r (g×10^{-3}/mol)	[14C]ATP MA (min^{-1})	[14C]ATP M_r (g×10^{-3}/mol)
95,000 to 101,000	38,000 to 40,000	1.99	0.80	179–534	3210	232	5720	274	3048	220
(α1) 98,900 (α2) 87,500	(β) 39,400			486–594	—	—	—	—	—	—

[a]Taken from Peterson *et al.* (1978b, 1982a) and Peterson and Hokin (1982).

TABLE XI

Amino Acid and Carbohydrate Composition of Brine
Shrimp Na$^+$,K$^+$-ATPase Subunits[a]

Residue	α-Subunit	β-Subunit
Amino acids		
Lysine	5.0	6.0
Histidine	1.0	0.8
Arginine	3.0	2.6
S-Sulfocysteine	2.7	3.2
Aspartic acid	10.4	11.0
Threonine	6.9	5.8
Serine	7.4	7.4
Glutamic acid	9.7	10.2
Proline	4.4	5.4
Glycine	7.6	8.2
Alanine	8.4	5.7
Valine	5.3	5.6
Methionine	3.3	2.6
Isoleucine	5.7	4.9
Leucine	9.0	7.6
Tyrosine	2.9	4.1
Phenylalanine	5.1	5.3
Tryptophan	2.1	3.6
Carbohydrates		
Amino sugars	0.66	1.06
Neutral sugars	2.9	5.7
Sialic acid	0	0
Total percentage by weight	5.20	9.19

[a]Taken from Peterson et al. (1982b) and Peterson and Hokin
(1980).

2. Cytochemical Localization of Na$^+$,K$^+$-ATPase: Isolated Larval Salt Gland and Midgut Intestinal Tissue

In light of these findings, it was most important to gather information about the intracellular localization of Na$^+$,K$^+$-ATPase in order to determine whether the two different kinds of α-subunits found in naupliar brine shrimp were derived from (1) a single cell type having two molecular forms on different spatial surfaces (i.e., apical versus basolateral) or (2) two cell types, each having a single molecular form which differed only in the carbohydrate moiety of the α-subunit.

Our earlier studies on quantitative localization of Na$^+$,K$^+$-ATPase activity have shown that the larval salt gland contains approximately 20% of the total enzymatic activity of the embryo whereas the midgut intestine contains between 25 and 45% (Ewing *et al.*, 1974; Lowy, 1984). The Na^+,K^+-ATPase activity in these tissues can be maintained for long periods of time in a simple basal salt culture medium (SGBM) and if the SGBM were provided with appropriate substrates, trapping agents, buffer fixatives, and inhibitors, one could prepare a type of cytochemical reaction mixture suitable for the intracellular localization of the Na^+,K^+-ATPase (Ernst, 1972a,b). It was necessary to make modification of the K^+-NPPase cytochemical technique because it has been reported that the substrate *p*-nitrophenylphosphate (NPP) is readily hydrolyzed by alkaline phosphatase (AP) at high pH (Ernst and Hootman, 1981). Previously, we reported finding abundant AP in naupliar brine shrimp plasma membranes (Peterson *et al.*, 1978b) and it was obvious that a blocking agent must be added to the cytochemical mixture in order to distinguish between the release of phosphate ion (P_i) from NPP substrate by Na^+,K^+-ATPase rather than the nonspecific AP. One of the most effective inhibitory agents of AP is the family of antihelminthic drugs, tetramisole or its analog (levamisole) (Borgers, 1973; Firth and Marland, 1975). If one chose not to use these inhibitors, then AP can be differentiated from Na^+,K^+-ATPase by its K^+ independence and ouabain insensitivity (Ernst, 1975). In developing our protocol we chose to utilize all three precautions.

The K^+-NPPase cytochemical technique employed was modified in the following manner: (1) lead citrate was substituted for strontium as the capture reagent, (2) dimethyl sulfoxide (DMSO) was included in the reaction mixture to enhance K^+-NPPase activity (Mayahara *et al.*, 1981), and (3) incubation times in the cytochemical reaction medium were extended from 1.5 to 3 hours.

The standard culture media was removed from the tissues by washing ($2\times$) with cold-buffered cacodylate (120 mM), pH 7.2, then stabilized in a cold cacodylate-buffered solution containing 2% formaldehyde plus 0.5% glutaraldehyde for 1 hour. Following stabilization, tissues were washed ($2\times$) with a cold 250 mM sucrose–cacodylate buffer solution containing 10% DMSO and stored in the cold (4°C) to await being used in the enzyme incubation medium. The K^+-NPPase incubation medium was freshly prepared. A complete reaction mixture contained the following: 250 mM glycine–KOH buffer, pH 8.8, 4 mM lead citrate, 25% DMSO, 10 mM *p*-nitrophenylphosphate, 10 mM Mg^{2+}, 45 mM K^+, and 2.5 mM levamisole. NaOH was substituted for KOH in preparing K-deficient reagents. After incubation, the tissue was washed ($2\times$) with 100 mM cacodylate buffer, pH 7.2, containing 8% sucrose and postfixed for 0.5 hour in 1% OsO_4–4% sucrose–100 mM cacodylate buffer, pH 7.2, solution. After two distilled water rinses, the tissue was dehydrated, embedded, and sectioned for TEM (Hootman and Conte, 1975.

Figure 19A, B, and C shows the results obtained when the salt gland was placed in the reaction mixture that was complete except for (1) absence of substrate NPP, (2) absence of K^+, and (3) addition of 10 mM ouabain. None of the outer or inner cellular membranes depicts phosphate reaction product, indicating that AP and Na^+,K^+-ATPase activities were blocked. Figure 19D,E is the electron photomicrograph of the crustacean chloride cell treated with complete reaction mixture containing 2.5 mM levamisole. Most localization of K^+-NPPase activity is found in the plasma membranes of tubular labyrinth; however, there is activity found in plasma membranes surrounding yolk platelets and juxtaposed to mitochondria. Plasma membrane projections lying beneath the endocuticle show appreciable activity as seen in Fig. 19F, indicating that K^+-NPPase activity is present in the apical tufts. Membranes of the nuclear envelope, RER, and SER are devoid of K^+-NPPase activity. Figure 20A is a low-magnification electron photomicrograph of the midgut intestine showing the anatomical area where the ion absorptive cell is found. The ultrastructural features of the ion absorptive cell are very clear especially the numerous microvilli (Fig. 20B). When isolated midgut intestine was placed in the "control" type incubation mixtures, the ion absorptive cells were devoid of K^+-NPPase activity as shown in Fig. 20B. If one compares the ion absorptive cell from complete K^+-NPPase reaction mixture containing 2.5 mM levamisole (Fig. 20C) to the control (Fig. 20B), one observes that the K^+-NNPase activity is localized to the plasma membrane of the basolateral surfaces and microvillar border. Membranes surrounding the nucleus, mitochondria, RER, and SER are devoid of enzymatic activity.

In summary, comparison of the EM photomicrographs of the crustacean chloride cell and ion-absorptive cell supports an interpretation that the nonspecific AP has been inhibited by levamisole and the K^+-NPPase activity found is K^+ dependent and ouabain sensitive. Therefore the K^+-NPPase is revealing the location of the microdomains of Na^+,K^+-ATPase in the plasma membrane. Unfortunately, the unusual localization of Na^+,K^+-ATPase seen in the apical tufts of the crustacean chloride cell, which might offer a subcellular mechanism for the electrophysiological evidence of a small component of active sodium transport across the salt gland, is compromised by the evidence suggesting that Na^+,K^+-ATPase membranes are also found surrounding yolk platelets and mitochondrial complexes. These latter findings, at present, are without explanation despite the safeguards taken. Since the crustacean chloride cell and ion absorptive cell also demonstrate the presence of large quantities of Na^+,K^+-ATPase existing in the basolateral surfaces where it has been previously localized in vertebrate epithelial cells, we are certain that Na^+,K^+-ATPase will be found in apical plasma membrane. However, one should remain cautious about our findings and await further evidence (Kraehenbuhl et al., 1983).

The cytochemical localization of Na^+,K^+-ATPase in these two types of epithelial tissue lends support to the concept that two different cells could have distinct molecular forms. Until batch preparations of the two organs have had the

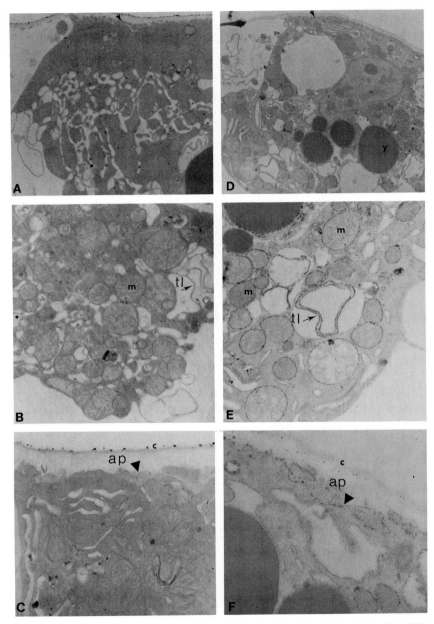

Fig. 19. TEM photomicrographs of isolated salt gland cells immersed in K^+-NPPase cytochemical reaction mixture described in text. (A) Gland incubated in absence of substrate, NPP ($\times 10{,}700$); (B) gland incubated in absence of K^+ ion ($\times 12{,}800$); (C) Gland incubated in 10 mM

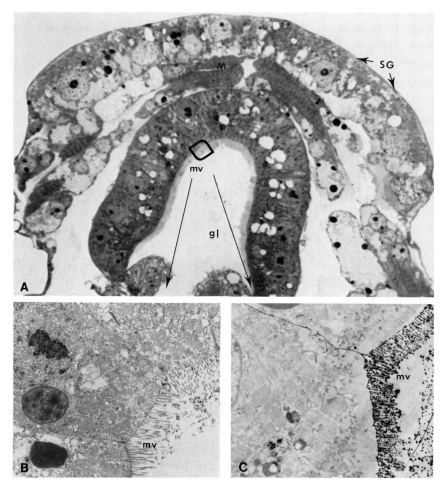

FIG. 20. TEM photomicrograph of isolated mid-gut ion absorptive cells immersed in K$^+$-NP-Pase cytochemical reaction mixture. (A) Transverse section of mid-gut showing numerous microvillii (mv) projecting into gut lumen (gl). Salt gland (SG) and swimming antennal muscles (M) lie in hemocoelomic cavity above the gut. (×1200); (B) mid-gut incubated in the absence of substrate, NPP (×19,800); (C) mid-gut in a complete reaction mixture (×19,800).

ouabain (×19,400); (D) gland incubated in complete reaction mixture showing P$_i$ deposits at arrow located at the apical surface. Note yolk platelet (y) (×10,000); (E) basolateral surface of chloride cell showing large P$_i$ deposits in the tubular labyrinth (tl) membranes of central cytoplasmic zone. Note deposits of P$_i$ around yolk platelets and mitochondria (m) ×20,000); (F) apical surface, at higher magnification, showing P$_i$ deposits on and around tubular tuft (ap) projections (×37,800).

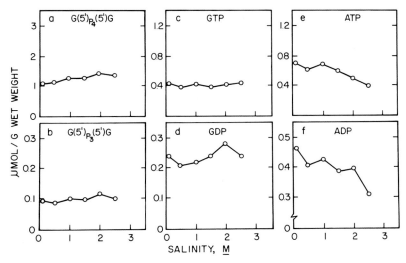

F<small>IG</small>. 21.   (a–f) Effects of salinity upon free adenine and guanine nucleotides and nucleotide storage compounds found in yolk platelets. Courtesy of Warner and Morrison, Univ. of Windsor, Ontario, Canada.

individual Na^+,K^+-ATPases isolated, purified, and SDS–PAGE analyzed, the concept remains unresolved, as does the concept of a single cell having two different molecular forms on various plasma membrane domains. Until special molecular probes can be made from antisera to α-subunits in which one could distinguish between α1- and α2-subunits one cannot determine if the Na^+,K^+-ATPase spatial orientation observed in the crustacean chloride cell is due to different α-subunit epitopes.

VI. Bioenergetics in Larval Brine Shrimp

One of the basic tenets of intermediary metabolism in animal cells is that for each metabolic sequence that uses or regenerates ATP, the overall balance of the cellular energy system must depend in large part on the stoichiometries of ATP consumption and production.

We have shown, rather conclusively, that the larval salt gland and midgut intestine contain most of the naupliar Na^+,K^+-ATPase. Therefore, any movement of Na^+ and K^+ across the membrane boundaries of these cells that utilizes this enzyme will place a notable demand upon the cellular high-energy nucleotide pool. How does the crustacean chloride cell (or for that matter the nauplius itself) cope with this ATP utilization. To answer this question we have to obtain direct experimental evidence on the intracellular metabolic pathways that control cation and anion movement across the interior of individual epithelial cells. Unfortunately, that information is unavailable but we do have a great amount of experimental information dealing with heterogeneity of the Na^+, K^+-ATPase-linked metabolic pathways in the intact nauplius. Therefore, let us review the evidence on aerobic and anaerobic metabolism in brine shrimp to gain insight regarding this problem (Clegg and Conte, 1980).

A. Aerobic Metabolism: Oxygen Consumption, Carbohydrate, Lipid, and Nucleotide Metabolism

1. Adult Brine Shrimp

Respiratory rates for adult brine shrimp have been measured in different environmental oxygen and salt concentrations. In a constant salinity (35‰), but varied concentrations of environmental oxygen, Decleir et al. (1980) showed that the rate of respiration for males was maintained at 2.7 µl O_2/individual/hour except when animals approached a critical oxygen tension zone (2.0 ml O_2/ml). Below this oxygen concentration the respiratory rate diminished precipitously to 0.7 µl O_2/individual/hour and animals began to excrete large amounts of different types of anaerobic metabolic end-products. Salinity effects on the respiratory rates of adult males and females have been studied by Gilchrist (1956, 1958) who found little change in respiratory rates (approx. 2.3 µl O_2/individual/hour) between salinities of 35 to 140‰. Using this respiratory data together with ion flux measurements, Croghan (1958e) calculated the minimum thermodynamic expenditure needed to maintain water and electrolyte balance. He found that only 6% of the total available metabolic energy is needed for osmotic work in adult tissues and does not constitute a heavy drain on the high-energy nucleotide pool. Unfortunately, measurements of nucleotide concentrations after acclimation to various saline gradients have not been measured in adult shrimp.

2. Naupliar Brine Shrimp

Respiratory rates for nauplii have been measured in different salt and oxygen concentrations (Conte et al., 1980). They are similar to the findings for adults with measured values being nearly constant (ranging from 12.0 to 16.0 µl O_2/hour/mg protein) for animals living in salinities going from salt concentrations of $0.05 \rightarrow 2.5$ M NaCl with oxygen tension values equal to or greater than the critical oxygen tension of 2.0 ml O_2/liter. Therefore, when nauplii are at high salinities and contain sufficient dissolved oxygen (> 2 ml O_2/liter), the cytoplasmic levels of high-energy nucleotides should be maintained because naupliar tissues are being furnished with an adequate supply of gaseous oxygen to sustain oxidative phosphorylation. Figure 21 contains the measured concentrations of free nucleotides and nucleotide storage compounds isolated from larval tissue following a 6-hour acclimation of the nauplii to various oxygenated salinities. Respiratory rates of these nauplii were in the range of 12 μl O_2/hour/mg protein. It can be seen from the data that there is a marked reduction in ATP and ADP concentrations while there is an increase in GDP concentration at higher salinities (>1.0 M NaCl) despite the constant oxygen utilization. GTP and the yolk storage compounds (Gp_3G and Gp_4G) concentrations appear to remain unchanged. Apparently, unlike adult ion transporting tissues, a situation exists in the larval tissues where aerobic oxidation–reduction reactions are either (1) being inhibited by microenvironmental changes in ionic composition or (2) being maintained at a maximal rate of ATP production, which becomes insufficient to meet the increased cellular ATP utilization stimulated by the steep salt gradient. When a situation of this nature occurs, it is usually assumed that facultative anaerobic pathways will be invoked to compensate for the declining levels of ATP. One such situation is most illustrative in this regard, and that is when environmental oxygen is absent (anoxia). Then glycolysis becomes the predominate bioenergetic pathway and lactate the major metabolic end-product. In anoxic nauplii, production of large amounts of cytoplasmic lactate together with increase excretion of lactic acid into the external medium have been reported (Ewing and Clegg, 1969) which indicates that under anoxic conditions nauplii utilize the typical glycolytic–lactate pathway.

Table XII contains results that were taken from oxygenated nauplii acclimated to two different NaCl concentrations. Each group was assayed for glycogen, lactate, oxygen consumption, acid production (pH), fatty acids, and adenine nucleotides. Despite the presence of large amounts of dissolved oxygen and showing a constant respiratory rate, nauplii evidenced a loss of glycogen and ATP at the higher salinity. At the same time, the acidity of the medium increased whereas the level of lactate and lipids remained unchanged. It appears the increased NaCl gradient causes stimulation of glycolysis, but the presence of oxygen prevents synthesis of lactate. Is the salt-dependent glycolysis caused by the stimulation of Na^+,K^+-ATPase?

B. Aerobic Metabolism: Sodium-Dependent Glycolysis and CO_2 Fixation

1. Na^+,K^+-ATPase-Coupled Glycolysis

If the sodium ion contained in the saline is responsible for stimulation of glycolysis, then inhibition of the Na^+,K^+-ATPase mediated transport of sodium ions across the plasma membrane should prevent glycogen utilization. Table XIII is a comparison of glycolysis and lactic acid production in nauplii acclimated to low and high salt concentrations in the presence and absence of ouabain. The

TABLE XII

The Effect of a NaCl Gradient on Carbohydrate, Lipid, and Intermediate Metabolism End-Products

External media (M. NaCl)	Glycolysis[a] (µg glucose/ hour/mg protein)	Acidification of media (ΔpH/hour/mg)	Rate of respiration[b] (µl/O_2/hour/mg)	Lactate production[a] (µg/hour/mg)	Lipolysis:[a] esterified fatty acids (ng/hour/mg)	ATP content (µmol/g)	ADP content (µmol/g)
0.5	8.3 ± 0.1[c]	0.0012[c]	12.4 ± 1.0	<1.00	1.02 ± 0.01	1.12 ± 0.01[c]	0.24 ± 0.01
2.5	13.3 ± 0.1[c]	0.0040[c]	11.6 ± 1.0	<1.00	1.01 ± 0.01	0.88 ± 0.02[c]	0.22 ± 0.01

[a]Rate of production or breakdown was derived from measurements taken over a 6-hour time period; 24-hour-old nauplii were used in all experiments.
[b]Pure oxygen was used to saturate the media. Dissolved oxygen in media was equal to or greater than 2.5 ml O_2/liter.
[c]Significant differences exist between the mean values using Student's t test ($p < 0.001$).

TABLE XIII

COMPARISON OF GLYCOGEN UTILIZATION AND LACTIC ACID PRODUCTION IN NAUPLII DURING
Na^+,K^+-ATPase INHIBITION UNDER TWO DIFFERENT AEROBIC SALINITIES[a]

Salinity (M NaCl)	Hours of incubation	Glycogen decrease[b] (μg glucose/mg protein)	Lactic acid[c] increase (μg/mg protein)
0.5	6	44	<1
0.5 plus $1 \times 10^{-4}\ M$ ouabain	6	35	<1
0.5 plus $4 \times 10^{-4}\ M$ ouabain	6	25	<1
2.5	6	80	<1
2.5 plus $1 \times 10^{-4}\ M$ ouabain	6	69	<1
2.5 plus $4 \times 10^{-4}\ M$ ouabain	6	48	<1

[a]Pure oxygen was used to saturate the media. Oxygen consumption for nauplii in both salinities was between 14 and 16 μl/hour/mg. Dissolved oxygen in media was equal to or greater than 2.5 ml O_2/liter.

[b]Initial (0 hr) glycogen content of nauplii minus final (6 hours) glycogen content of nauplii.

[c]Production of lactate is derived from initial (0 hour) lactate content of nauplii minus final (6 hour) lactate content of nauplii.

data show that addition to $10^{-4}\ M$ ouabain to the medium caused a reduction in glycolysis with no change in lactate production. Also, we have observed (unpublished) that ouabain is without effect upon oxygen consumption in salt-acclimated nauplii. From these results, we concluded that, in naupliar tissue, the enzyme Na^+,K^+-ATPase, performs a dual function: (1) regulates water and electrolyte balance by maintaining a transepithelial electrochemical gradient and (2) regulates cellular high energy nucleotide pooly by controlling intracellular glycogen breakdown.

2. Ion Coupled C-4 Organic Acid Facultative Pathway

The stimulation of glycogenolysis of Na^+,K^+-ATPase cannot cause a continuous breakdown of glucose unless there is available an adequate amount of cofactor (NAD^+). Normally, lactate production would provide the cytoplasmic NAD^+ during anoxic conditions but the presence of oxygen interferes with this metabolic pathway. Since the stimulatory effect of Na^+,K^+-ATPase does not appear directly linked to oxidative metabolism for supplying NAD^+, some other nonmitochondrial source of NAD^+ must be utilized. I proposed (Conte, 1980) that a C-4 dicarboxylic acid pathway could serve as a facultative shunt enabling a shuttle between oxaloacetate and malate to furnish the needed cytoplasmic NAD^+. This mechanism requires the fixation of CO_2 from exogenous bicarbonate to form oxaloacetate from phosphenolypyruvate and results in (1) formation of dicarboxylic acids and/or amino acids via transamination and (2) reduction of oxaloacetate to malate by oxidation–reduction of the cofactor NADH (NADH $\rightarrow$ NAD^+).

Experimental evidence which supports the existence of the facultative shunt in larval brine shrimp is summarized as follows.

a. *MDH and PEPCK.* Two key enzymes, which are necessary for the existence of the facultative shunt, malate dehydrogenase (MDH) and phosphoenolpyruvate carboxykinase (PEPCK) have been found in rather substantial quantities in naupliar tissue. They have been isolated, characterized, and partially purified (Table XIV).

b. *sMDH Regulation.* Recall that during early larval development we found a NaCl enhancement of Na^+,K^+-ATPase biosynthesis synchronized to naupliar sodium regulation (Conte *et al.,* 1977). Likewise, a comparison of larval MDH biosynthetic rate and developmental pattern in nauplii exposed to elevated salinities showed the cytosolic-derived enzyme (sMDH) rather than the mitochondrial-derived enzyme (sMDH) to be preferentially synthesized at higher rates. We interpreted these results of a NaCl-stimulated sMDH activity as a reflection of an enrichment of the oxidoreductase capabilities needed by the embryo during periods of high glucose metabolism coupled with active transport of sodium (Hand *et al.,* 1981; Hand and Conte, 1982a,b).

TABLE XIV

Yield of Na^+,K^+-Activated ATPase, Malate Dehydrogenase, and Phosphoenolypyruvate Carboxykinase from 200 g Wet Weight of Brine Shrimp Nauplii

Enzyme	Total		Specific activity (units/mg prot.)	Specific activity at highest state of purity (units/mg prot.)
	Protein	Units		
Na^+,K^+-activated ATPase[a]	14,600	28,200	1.94	514
Malate dehydrogenase[b]	2,560	18,061	7.06	570
Phosphoenolpyruvate carboxykinase[c]	2,110	2,210	0.95	170

[a] Units of activity are expressed in µmol P_i/hour for whole homogenates. See Peterson *et al.* (1978a, 1982a,b) for isolation, characterization, and purification.

[b] Units of activity are expressed in micromoles. NADH oxidized/hour for postmitochondrial supernatant. See Hand and Conte (1982a,b) for isolation, characterization, and purification.

[c] Units of activity are expressed in nanomoles $H^{14}CO_3$ incorporates/minute for postmitochondrial supernatant. Isolation, characterization, and purification are unpublished at this time.

c. *Release and Excretion of Ammonia and Free Amino Acids.* Recently, we studied the effect of increased environmental salinities on ammonia excretion and the metabolic pathways that release free amino acids into the intracellular compartment, either by the breakdown of storage proteins in yolk platelets (Warner *et al.*, 1972) or by biosynthetic reactions. The results are shown in Table XV; it can be seen that the intracellular free amino acids rise markedly in response to the NaCl gradient and ammonia excretion has doubled. Amino acid excretion appears to increased slightly which is similar to that found by Emerson (1967) while investigating free amino acid metabolism in developing brine shrimp embryos.

d. *Formation of Radioactively Labeled Malate and Five Amino Acids.* Clegg (1976), investigating interrelationships between water and cellular metabolism, found $^{14}CO_2$ to be readily incorporated into dicarboxylic acids, amino acids, and pyrimidines. Among the dicarboxylic acids, most of the radioactivity was found in malate (8–10%). During our investigation of the biosynthesis of ^{14}C-labeled amino acids from the incorporation of $^{14}CO_2$ (as $H^{14}CO_3$), we (Conte and Carpenter, 1984) found the amino acids contained 50–60% of the total radioactivity similar to that found by Clegg (1976) and Peterson *et al.* (1982a) for low salinity (0.5 M NaCl). However, by simply increasing the salt gradient to very high levels (>1.5 M NaCl), it was possible to raise the [^{14}C]malate concentration nearly 3-fold (25–30%) and elevate the amino acids to 70% of the total radioactivity. In addition, among the 20 plus free amino acids found in the intracellular pool, only five amino acids, namely aspartate, glutamate, serine, proline, and alanine, contained the ^{14}C label. The stimulation of the $^{14}CO_2$ amino acid biosynthetic pathways by elevation of the environmental salt concentration did not cause an increase in the specific activity for any of the five ^{14}C-labeled amino acids. Table XVI shows the results of this experiment and it can be seen that alanine and proline concentrations have risen while aspartate and glutamate concentration diminished and serine remained constant. No explanation can be offered at this time for these shifts between amino acids. However, the NaCl gradient which effectively causes stimulation of the glycogenolytic and glycolytic pathways does appear to have a similar effect upon the CO_2 fixation pathway.

C. REGULATION OF SODIUM-DEPENDENT GLYCOLYSIS: STIMULATION OF PHOSPHODIESTERASE OR INHIBITION OF ORGANIC ANION TRANSPORT

As was disclosed in Section II,C,4, we found nauplii to be very sensitive to sulfonamides with the toxicity exhibited by the larvae to be salt dependent. Unsubstituted sulfonamides (R-SO$_2$ NH$_2$) are specific inhibitors of carbonic anhydrase and the physiological responses to these substances have up to now been interpreted as consequences of CA inhibition (Maren, 1977; Maren and

TABLE XV

The Effect of NaCl Gradient on Ammonia Excretion and Intracellular Free Amino Acids under[a] Aerobic Conditions

Salinity[b] in (M, NaCl)	Total ammonia in cytosol (μmol/100 mg prot.)	Total ammonia in external medium (μmol/100 mg prot.)	Total amino acids in cytosol (μmol/100 mg prot.)	Total amino acids in external medium (μmol/100 mg prot.)
0.25	1.90 ± 0.01[c]	2.60 ± 0.08	22.40 ± 0.06	1.40 ± 0.08
0.50	2.00 ± 0.01	2.50 ± 0.06	25.70 ± 0.08	2.60 ± 0.07
1.00	2.10 ± 0.06	3.20 ± 0.07	26.10 ± 0.03	2.10 ± 0.09
1.50	3.40 ± 0.02	4.80 ± 0.07	30.50 ± 0.09	3.30 ± 0.02
2.00	2.70 ± 0.02	6.50 ± 0.08	41.10 ± 0.06	2.10 ± 0.05
2.50	3.90 ± 0.05	4.30 ± 0.05	45.90 ± 0.03	3.30 ± 0.08

[a] Pure oxygen used to saturate the media. Dissolved oxygen ranged from 12.8 to 4.3 ml O_2/liter. Oxygen consumption ranged from 12 to 14 μl O_2/hour/mg prot.

[b] NaCl gradient prepared by adding NaCl to sterilized acclimation medium containing 10 mM K$^+$, 6 mM Ca^{2+}, 6 mM Mg^{2+}, 3 mM HCO$_3$ at pH 7.8 plus penicillin–streptomycin. Naupliar densities equal to 0.6–0.8 g nauplii per 20 ml of acclimation medium.

[c] Mean value $\pm$ 1 SD.

TABLE XVI

The Effect of NaCl Gradient on Glycolysis and the Formation of ^{14}C-Labeled Amino Acids via CO_2 Fixation Pathways in Naupliar Cytosol under Aerobic Conditions[a]

Salinity[b] in (M/NaCl)	Net glycolysis[c] (μg glucose/ mg prot.)	Percentage ^{14}C-labeled amino acids[d]					
		Aspartate (% total cpm/ mg prot.)	Glutamate (% total cpm/mg prot.)	Serine (% total cpm/ mg prot.)	Proline (% total cpm/ mg prot.)	Alanine (% total cpm/ mg prot.)
0.25	17	30.2	26.7	8.6	14.4	9.0
0.50	20	30.7	33.4	7.3	14.8	8.2
1.00	24	29.4	25.9	8.8	17.9	10.7
1.50	49	26.2	28.0	6.8	20.8	10.6
2.00	54	23.1	23.3	9.4	20.9	20.9
2.50	118	21.5	18.3	7.3	20.9	28.1

[a]Pure oxygen was used to saturate the media. Dissolved oxygen ranged from 1.8 to 4.3 ml O_2/liter. Oxygen consumption ranged from 12 to 14 μl O_2/hour/mg protein.

[b]NaCl gradient was prepared by adding NaCl to sterilized acclimation media containing 10 mM K$^+$, 6 mM Ca^{2+}, 6 mM Mg^{2+}, 3 mM HCO_3 at pH 7.8 plus penicillin–streptomycin.

[c]Net glycolysis is equal to initial glycogen content minus final glycogen content. Initial glycogen content for newly hatched nauplii in 0.5 M NaCl antibiotic artificial seawater medium prior to transfer into NaCl fortified acclimation media was 142 ± 1 μg glucose/mg prot.

[d]Conditioned media did not contain any ^{14}C-labeled amino acids.

Sanyal, 1983). Evidence has been accumulating which indicates that inhibitory action of sulfonamide is not limited to CA but can effect other enzymes, such as alkaline phosphatase (AP). This is not surprising since AP is another zinc-containing metalloprotein like CA (Price, 1979). Recall that our naupliar preparation contained miniscule amounts of CA and large amounts of AP (Peterson *et al.*, 1978a) and therefore we could not discern if these two enzymes were key elements in the observed naupliar toxicity. In these circumstances, we thought it worthwhile to compare the action of acetazolamide to the alkylated compound, *N-t*-butylacetazolamide. This comparison differentiates between the effects of zinc-containing enzymes to non-zinc-containing proteins upon naupliar metabolism. The results obtained from nauplii acclimated for 3 hours in fully oxygenated, NaCl-enriched salines containing either 1 mM acetazolamide or 1 mM *t*-butylacetazolamide are shown in Table XVII. Nauplii were assayed for glycogen, ammonia excretion, free amino acids, and ^{14}C-labeled amino acids. As one can see from the results, each compound stimulated glycolysis, proteolysis, and $^{14}CO_2$ fixation. It is unlikely that inhibition of CA or AP can account for these effects due to the fact that the akylated derivative is as effective as the parent compound. Therefore, the unsubstituted amide moiety of the sulfonamide is not the site of action for this phenomenon. One must consider other aspects of the molecular structure between these two compounds to account for their mechanism of action. Since the drugs are interfering with glycolysis, one might suspect that cyclic nucleotide regulation is being disrupted. Schultz and Senft (1967) have reported that acetazolamide inhibits phosphodiesterase and Cousin and Motais (1976) showed that several types of alkyl-substituted sulfonamides (including *N-t*-butylacetazolamide) block anion transport, mainly organic anions and chloride. Since [^{14}C]bicarbonate uptake and incorporation were stimulated and amino acid extrusion appeared functional, it seems unlikely that organic anion transport is inhibited. It is more likely that these two sulfonamides which contain the thiadiazole ring are acting as stimulators to phosphodiesterase much in the same fashion as reported for imidazole (Butcher and Sutherland, 1962; Amer and Kreighbaum, 1975). If this is the case, then the increases in glycogenolysis and glycolysis seen in the naupliar cytosol might well be due to high turnover rates for cyclic nucleotides (cAMP or cGMP) caused by the greater phosphodiesterase activity. What is the explanation for the signal to release cAMP or cGMP. Here we return to the mode of action of Na^+,K^+-ATPase where increased NaCl gradient causes stimulation of the cationic enzyme and results in an elevation of K^+ in the cytoplasm. In turn the high cytosolic K^+ could be the signal for adenylate cyclase to synthesize and release of cyclic nucleotides (Ross and Gilman, 1980). Unfortunately, cyclic nucleotide levels have not been measured in the larval brine shrimp under normal or hypersaline conditions.

TABLE XVII

Comparison of NaCl Gradient on Glycolysis, Ammonia Excretion, Intracellular Free Amino Acids, and the Formation of ^{14}C-Labeled Amino Acids in Naupliar Cytosol in the Presence of 10^{-3} M Acetazolamide and Its Alkylated Derivative[a]

Salinity in (M, NaCl)	Net glycolysis (μg glucose/ mg prot.)	Ammonia (μmol/mg prot.)	Total free amino acids in cytosol (μmol/mg prot.)	Percentage ^{14}C-labeled amino acids (% of total cpm/mg protein)				
				Aspartate	Glutamate	Serine	Proline	Alanine
0.5 control	20	3.8	27.6	16.1	17.5	3.8	7.7	4.3
0.5 plus acetazolamide	52	4.6	53.4	13.6	15.6	4.0	7.2	4.6
0.5 plus N-t-butylacetazolamide	54	5.1	65.9	20.0	12.8	4.2	3.8	6.3
2.5 control	107	6.6	43.5	13.3	11.3	4.4	12.9	17.4
2.5 plus acetazolamide	124	16.6	124.1	14.2	14.0	1.7	12.7	20.5
2.5 plus N-t-butylacetazolamide	128	7.1	94.2	21.2	21.6	6.4	9.9	16.5

[a]Experimental conditions similar to Table XVI.

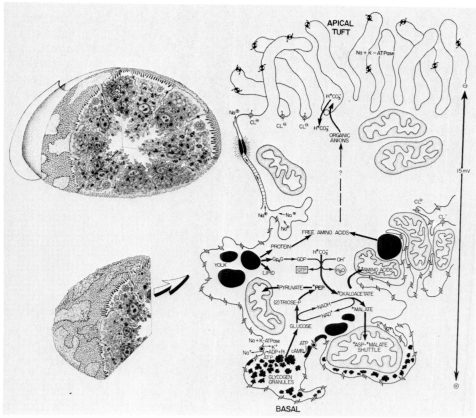

FIG. 22. Drawing of larval salt gland and cytodifferentiation of the chloride cell. Diagram depicts the molecular mechanisms involved in salt extrusion from the crustacean chloride cell.

VII. Model of Larval Salt Gland and Chloride Cell Differentiation

A. DEVELOPMENT OF LARVAL SALT GLAND

The cellular differentiation in the larval salt gland begins at the late prenaupliar stage with the formation of the cationic transport enzyme, Na$^+$,K$^+$-ATPase, from newly synthesized subunits (Fig. 22). The embryonic (or I should say undifferentiated) chloride cell is devoid of any extensive plasma membrane proliferation. However, this embryonic chloride cell contains large numbers of yolk platelets, many mitochondria, and numerous fields of glycogen granules located in the central cytoplasmic zone. Following activation of the Na$^+$,K$^+$-

ATPase genes, chloride cell differentiation begins in earnest with most of the cells in the gland exhibiting amplification of all plasma membrane surfaces including at basal, lateral, and apical boundaries. The mature larval chloride cell, which is considered to be functional, has two distinct allomorphic patterns of plasma membrane growth. First, the tubular labyrinth is derived from invagination of basolateral surfaces, is tubular in appearance, and is closely associated with mitochondria. Copeland (1966, 1967) refers to this mitochondrial–plasma membrane complex as being ''a metabolically linked ion pump.'' Second, the tubular tufts are outgrowths of the apical surface, are irregularly shaped, and are not associated with mitochondria. The larval salt gland is a transient structure. Beginning at the juvenile stage of metamorphosis, the mature chloride cell undergoes dedifferentiation and gland is reabsorped.

B. Cytomolecular Compartmentalization of the Chloride Cell

The bioenergetic pathways furnishing high-energy intermediates needed by the microdomains of the plasma membrane for active ion transport appear derived from both mitochondrial and extramitochondrial compartments.

1. Na^+,K^+-ATPase Linked Glycolysis–Vitellolysis

The disappearance of glycogen granules and yolk platelets in the chloride cell can now be explained by the effect of Na^+,K^+-ATPase (Fig. 23). The required ATP for the sodium pump is derived from both oxidative phosphorylation and the facultative anaerobic shunt. The breakdown of glycogen furnishes the needed substrate for both pathways, phosphoenolpyruvate (PEP). The C-4 dicarboxylic acid pathway serves to furnish the chloride cell with the cofactor NAD^+ (in the absence of oxygen) and produces GTP from GDP plus innocuous amino acids as end-products (instead of lactate). The amino acids synthesized from the CO_2 fixation pathway appear to be compartmentalized in the mitochondria. The breakdown of yolk platelets furnishes the cofactor GDP and provides for net amounts of high-energy GTP. Furthermore, the platelets contain the appropriate enzymes to recycle the GDP as depicted in the scheme in Fig. 23.

2. Na^+,K^+-ATPase Linked Adenylate Cyclase Complex

I speculated that the control of the energy flow between oxidative phosphorylation and the facultative anaerobic shunt is dependent upon cyclic nucleotides. If the epiforms of Na^+,K^+-ATPase, as found by the cytochemical localization, have different ligand binding for internal membrane components (such as for adenylate cyclase) then it is possible to consider that the Na^+,K^+-ATPase found in the tubular labyrinth having its ''link'' to glycolysis via a high affinity for adenylate cyclase.

If the apical tufts Na^+,K^+-ATPase had a low affinity to adenylate cyclase, then this epiform of the enzyme would not demonstrate a glycolytic effect. A

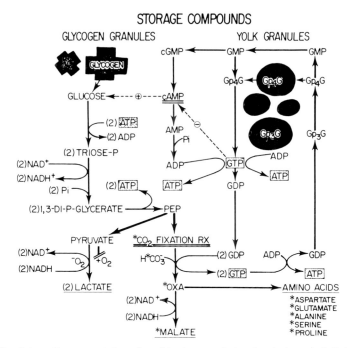

FIG. 23.   Schematic representation of aerobic glycogenolysis, glycolysis, and vitellolysis in larval brine shrimp.

great deal more work will have to be done on nucleotide metabolism and the purified α-subunits of the Na$^+$,K$^+$-ATPase to verify this point.

3. Na$^+$,K$^+$-ATPase Linked Anion Antiport System

The schematic shown in Fig. 22 depicts bicarbonate ion being transported into the apical region of the cell, possibly by an antiport system that exports organic acids, amino acids, or chloride ions. Once inside the cytoplasm, the bicarbonate is acted upon by PEPCK to form oxaloacetate and GPT with the possibility of forming metabolic water (OH$^-$ combining with the protons coming from the mitochondria). The oxaloacetate can then be reduced to malate or shuttled into the TCA cycle. The evidence to support and explain the details of the anion transport system is very meager and requires a great deal of additional research.

In summary, the role of the plasma membrane in the larval chloride cell is being unraveled. Through the use of new immunohistochemical techniques, one is hopeful that we can localize the precise distribution of Na$^+$,K$^+$-ATPase and other important enzyme complexes to explain the structure and function of the crustacean larval salt gland.

ACKNOWLEDGMENTS

The author expresses his deep appreciation to his colleagues, Drs. Seth R. Hootman, Richard D. Ewing, Gary L. Peterson, Steven Hand, Joel Lowy, and John Carpenter for the time spent in the laboratory and in the long and helpful discussions of our research. The author is also indebted to Ms. S. Sargent for manuscript preparation and Ms. Kathryn Torvik for preparation of the drawings. This research has been funded by National Science Foundation Grants PCM 76-14771 to F. P. Conte.

REFERENCES

Amer, M. S., and Kreighbaum, W. E. (1975). *J. Pharmacol. Sci.* **64,** 1–37.
Askarua, K. (1970). *Sci. Rep. Kanazawa Univ.* **15,** 37–56.
Baxter, L. A., Fisher, J. A., and Hokin, L. E. (1983). *J. Gen. Physiol.* **82,** 8a.
Borgers, M. (1973). *J. Histochem. Cytochem.* **21,** 812–824.
Bradley, T. J. (1983). *In* "Comprehensive Insect Physiology, Biochemistry and Pharmacology" (G. A. Kerkut and L. I. Gilbert, eds.), Ch. 14. Pergamon, Oxford.
Bradley, T. J., and Phillips, J. E. (1975). *J. Exp. Biol.* **63,** 331–342.
Bradley, T. J., and Phillips, J. E. (1977a). *J. Exp. Biol.* **66,** 83–96.
Bradley, T. J., and Phillips, J. E. (1977b). *J. Exp. Biol.* **66,** 97–110.
Bradley, R. J., and Phillips, J. E. (1977c). *J. Exp. Biol.* **66,** 111–126.
Bradley, T. J., Stuart, A. M., and Satir, P. (1982). *Tissue Cell* **14,** 759–773.
Burnett, L. E., Woodson, P. J., Rietow, M. G., and Vilrich, V. C. (1981). *J. Exp. Biol.* **92,** 243–254.
Butcher, R. W., and Sutherland, E. W. (1962). *J. Biol. Chem.* **237,** 1244–1250.
Carpenter, J. F., and Conte, F. P. (1982). *Am. Zool.* **22,** 896.
Churchill, L., Peterson, G. L., and Hokin, L. E. (1979). *Biochem. Biophys. Res. Commun.* **90,** 488–490.
Claus, C. (1876). *Z. Wiss. Zool.* **27,** 362–402.
Clegg, J. (1962). *Biol. Bull.* **123,** 295–301.
Clegg, J. (1964). *J. Exp. Biol.* **41,** 879–892.
Clegg, J. (1976). *J. Cell. Physiol.* **89,** 369–380.
Clegg, J., and Conte, F. P. (1980). *In* "The Brine Shrimp, *Artemia*" (G. Persoone, P. Sorgeloos, O. Roels, and E. Jaspers, eds.), Vol. II, pp. 1–54. Universa Press, Wetteren, Belgium.
Coleman, J. E. (1975). *In* "Inorganic Biochemistry" (G. L. Eichorn, ed.), Vols. 1 and 2, pp. 488–548. Elsevier, Amsterdam.
Conte, F. P. (1980). *Am. J. Physiol.* **238,** R269–R276.
Conte, F. P., and Carpenter (1984). *J. Exp. Zool.,* submitted.
Conte, F. P., and Hootman, S. R. (1974).
Conte, F. P., Hootman, S. R., and Harris, P. J. (1972). *J. Comp. Physiol.* **80,** 239–246.
Conte, F. P., Droukas, P. C., and Ewing, R. D. (1977). *J. Exp. Zool.* **202,** 339–362.
Conte, F. P., Lowy, J., Carpenter, J., Edwards, A., Smith, R., and Ewing, R. D. (1980). *In* "The Brine Shrimp, *Artemia*" (G. Persoone, P. Sorgeloos, O. Reols, and E. Jaspers, eds.), Vol. II, pp. 125–136. Universa Press, Wetteren, Belgium.
Copeland, D. E. (1966). *Science* **151,** 470–471.
Copeland, D. E. (1967). *Protoplasma* **63,** 363–384.
Cousin and Motais (1976).
Croghan, P. C. (1958a). *J. Exp. Biol.* **35,** 213–218.

Croghan, P. C. (1958b). *J. Exp. Biol.* **35**, 219–233.

Croghan, P. C. (1958c). *J. Exp. Biol.* **35**, 234–242.

Croghan, P. C. (1958d). *J. Exp. Biol.* **35**, 243–249.

Croghan, P. C. (1958e). *J. Exp. Biol.* **35**, 425–436.

Churchill, L., Peterson, G. L., and Hokin, L. E. (1979). *Biochem. Biophys. Res. Commun.* **90**, 488–490.

Decleir, W., Vos, J., Bernaerts, F., and Van den Branden, C. (1980). *In* "The Brine Shrimp" (G. Personne, P. Sorgeloos, O. Roels, and E. Jaspers, eds), Vol. II, pp. 137–145. Universa Press, Wetteren, Belgium.

Dejdar, E. (1930). *Z. Wiss. Zool.* **136**, 422–452.

Emerson, D. N. (1967). *Comp. Biochem. Physiol.* **20**, 245–261.

Ernst, S. A. (1972a). *J. Histochem. Cytochem.* **20**, 13–22.

Ernst, S. A. (1972b). *J. Histochem. Cytochem.* **20**, 23–38.

Ernst, S. A. (1975). *J. Cell Biol.* **66**, 586–608.

Ernst, S. A., and Hootman, S. R. (1981). *Histochem. J.* **13**, 397–418.

Ewing, R. D., and Clegg, J. (1969). *Comp. Biochem. Physiol.* **31**, 297–307.

Ewing, R. D., Peterson, G. L., and Conte, F. P. (1972). *J. Comp. Physiol.* **80**, 247–254.

Ewing, R. D., Peterson, G. L., and Conte, F. P. (1974). *J. Comp. Physiol.* **88**, 217–234.

Firth, J. A., and Marland, B. Y. (1975). *J. Histochem. Cytochem.* **23**, 571–574.

Geddes, M. C. (1975a). *Comp. Biochem. Physiol.* **51A**, 553–559.

Geddes, M. C. (1975b). *Comp. Biochem. Physiol.* **51A**, 561–571.

Geddes, M. C. (1975c). *Comp. Biochem. Physiol.* **51A**, 573–578.

Gilchrist, B. (1956). *Hydrobiologia* **8**, 54–60.

Gilchrist, B. (1958). *Hydrobiologia* **12**, 26–37.

Goldman, D. E. (1943). *J. Gen. Physiol.* **27**, 37–60.

Hand, S. C., and Conte, F. P. (1982a). *J. Exp. Zool.* **219**, 7–15.

Hand, S. C., and Conte, F. P. (1982b). *J. Exp. Zool.* **219**, 17–27.

Hand, S. C., Becker, M. M., and Conte, F. P. (1981). *J. Exp. Zool.* **217**, 199–323.

Heath, H. (1924). *J. Morphol.* **38**, 454–483.

Henry, R. P., and Cameron, J. N. (1982). *J. Exp. Zool.* **221**, 309–321.

Hers, H. G., and Hue, L. (1983). *Annu. Rev. Biochem.* **52**, 617–653.

Hootman, S. R., and Conte, F. P. (1974). *Cell Tissue Res.* **155**, 423–436.

Hootman, S. R., and Conte, F. P. (1975). *J. Morphol.* **145**, 371–386.

Hootman, S. R., Harris, P. J., and Conte, F. P. (1972). *J. Comp. Physiol.* **79**, 97–104.

Kikuchi, S. (1972). *Annu. Rep. Iwate Med. Univ.* **7**, 15–26.

Kirschner, L. B. (1977). *In* "Transport of Ions in Water in Animals" (B. Gupta, R. Moreton, J. Oschman, and B. Wall, eds.), pp. 427–452. Academic Press, New York.

Kirschner, L. B. (1979). *In* "Mechanism of Osmoregulation in Animals" (R. Gilles, ed.), pp. 157–222. Wiley, New York.

Kitahara, S., Fox, K. R., and Hogben, C.A.M. (1967). *Nature (London)* **214**, 836–837.

Komnick, H. (1977). *Int. Rev. Cytol.* **49**, 285–329.

Kraehenbuhl, J. P., Bonnard, C., Geering, K., Giradet, M., and Rossier, B. C. (1983). *J. Cell Biol.* **97**, 310A.

Leydig, F. (1851). *Z. Wiss. Zool.* **3**, 280–307.

Lowy, R. J. (1984). *Am. Zool.*, **24**, 265–274.

Lowy, R. J., and Conte, F. P. (1984). Submitted to *Am. J. Physiol.*

Lowy, R. J., Buckley, P. M., and Conte, F. P. (1981). *Cell Biol.* **91**, 250.

Maetz, J. (1971). *Philos. Trans. R. Soc. Ser. B* **262**, 209–249.

Maren, T. H. (1956). *J. Pharmacol. Exp. Ther.* **117**, 385–401.

Maren, T. H. (1967). *Physiol. Rev.* **47**, 597–765.

Maren, T. H. (1977). *Am. J. Physiol.* **232**, F291–F297.

Maren, T. H., and Sanyal, G. (1983). *Annu. Rev. Pharmacol. Toxicol.* **23**, 439–459.

Mayahara, H., Fujimoto, A. T., and Ogawa, K. (1981). *Histochemistry* **67**, 125–138.

Meredith, J., and Phillips, J. E. (1973a). *Can. J. Zool.* **51**, 349–353.

Meredith, J., and Phillips, J. E. (1973b). *J. Insect Physiol.* **19**, 1157–1172.

Peterson, G. L., and Hokin, L. E. (1980). *Biochem. J.* **192**, 107–118.

Peterson, G. L., and Hokin, L. E. (1982). *J. Biol. Chem.* **256**, 3751–3761.

Peterson, G. L., Ewing, R. D., and Conte, F. P. (1978a). *Dev. Biol.* **67**, 90–98.

Peterson, G. L., Ewing, R. D., Hootman, S. R., and Conte, F. P. (1978b). *J. Biol. Chem.* **253**, 4762–4770.

Peterson, G. L., Churchill, L., Fisher, J. A., and Hokin, L. E. (1982a). *J. Exp. Zool.* **221**, 295–308.

Peterson, G. L., Churchill, L., Fisher, J. A., and Hokin, L. E. (1982b). *Ann. N.Y. Acad. Sci.* **402**, 185–206.

Phillips, J. E. (1982). *Fed. Proc. Fed. Am. Soc. Exp. Biol.* **41**, 2348–2354.

Phillips, J. E., and Bradley, T. J. (1977). *In* "Transport of Ions and Water in Animal Tissues" (B. Gupta, J. Oschman, and B. Wall, eds.), pp. 709–732. Academic Press, New York.

Phillips, J. E., and Meredith, J. (1969). *Nature (London)* **222**, 168–169.

Potts, W. T. W., and Parry, G. (1984). *In* "Osmotic and Ionic Regulation in Animals," p. 319. Macmillan, New York.

Price, G. H. (1979). *Chem. Acta* **94**, 211–217.

Ross, E. M., and Gilman, A. G. (1980). *Annu. Rev. Biochem.* **49**, 533–564.

Russler, D., and Mangos, J. (1978). *Am. J. Physiol.* **234**, R216–R222.

Schultz, G., and Senft, G. (1967). *Arch. Exp. Pathol. Pharmakol.* **257**, 61–62.

Scott, W. N., and Skipski, I. (1979). *Comp. Biochem. Physiol.* **63B**, 429–435.

Scott, W. N., Shamoo, Y. E., and Brodsky, W. A. (1970). *Biochim. Biophys. Acta* **219**, 248–250.

Silverman, D. N., and Tu, C. K. (1976). *J. Am. Chem. Soc.* **98**, 478–524.

Smith, P. G. (1969a). *J. Exp. Biol.* **51**, 727–738.

Smith, P. G. (1969b). *J. Exp. Biol.* **51**, 739–757.

Spangenberg, F. (1875). *Z. Wiss. Zool.* **25** (Suppl.), 1–64.

Sweadner, K. J. (1979). *J. Biol. Chem.* **254**, 6060–6067.

Thuet, P., Motais, R., and Maetz, J. (1968). *Comp. Biochem. Physiol.* **26**, 793–818.

Urakabe, S., Shirai, D., Yusa, S., Kimura, S., Orita, Y., and Abe, H. (1976). *Comp. Biochem. Physiol.* **53C**, 115–119.

Ussing, H. H. (1949). *Acta Physiol. Scand.* **19**, 43–56.

Warner, A. H., Puodziukas, J. G., and Finamore, F. J. (1972). *Exp. Cell Res.* **70**, 365–375.

Warren, H. S. (1938). *J. Morphol.* **62**, 263–289.

Weisz, P. B. (1947). *J. Morphol.* **81**, 45–96.

Yancey, P. H., Clark, M. E., Hand, S. C., Bowlus, E. D., and Somero, G. N. (1982). *Science* **217**, 1214–1222.

Zenker, W. (1851). *Arch. Anat. Physiol. Med.* **1851**, 112–121.

Zograf, N. (1905). *C.R. Acad. Sci. (Paris)* **141**, 903–905.

Pathways of Endocytosis in Thyroid Follicle Cells

VOLKER HERZOG

*Department of Cell Biology, University of Munich, Munich,
Federal Republic of Germany*

I. Introduction

The thyroid gland maintains remarkably constant levels of thyroid hormones, thyroxine and triiodo thyronine, in the circulation of most mammalian species (Rall, 1974). This ability of the thyroid gland is based on the highly regulated sequence of steps in the secretory process: Thyroglobulin, the macromolecular secretory product, and the precursor of thyroid hormones, is released and stored in the follicle lumen where it remains segregated from the extrafollicular space by a tight epithelial monolayer. Under normal conditions, endocytosis is the only way by which thyroglobulin can be removed from the closed follicular lumen and by which it gains access to the proteolytic enzymes in lysosomes. The transfer of thyroglobulin from the follicle lumen to lysosomes is the main pathway of endocytosis in thyroid follicle cells and the prerequisite for the liberation of thyroid hormones (Wollman, 1969; Fujita, 1975; Lissitzky, 1976).

During endocytosis, the apical plasma membrane of thyroid follicle cells is

107

internalized approximately four times every hour. This amount corresponds roughly to the membrane area inserted during the release of thyroglobulin by exocytosis (Ericson, 1981). In the light of the previously observed membrane reutilization in secretory cells (Farquhar, 1981; Herzog, 1981; Steinman et al., 1983), it appears likely that at least part of the internalized membrane is not degraded in lysosomes but recycled through the follicle cell. Two pathways by which endocytic membrane may escape lysosomal degradation have been described.

1. Observations with electron-dense tracers suggest that part of the endocytic membranes detaches after fusion with endosomes and lysosomes and becomes inserted into the membranes of stacked Golgi cisternae (Herzog and Miller, 1979a, 1981).
2. A transepithelial vesicular transport operates in thyroid follicle cells with endocytic vesicle formation at the apical and exocytotic insertion into the basolateral cell surfaces. This vesicular transport across thyroid follicle cells includes the transfer of thyroglobulin from the follicle lumen to the extra-follicular space and explains its appearance in the circulation (Herzog, 1983a,b).

Studies on endocytosis in thyroid follicle cells require experimental access to the luminal cell surface. This has been accomplished in the past by microinjec-tion of tracers into the follicle lumen (Seljelid et al., 1970). More recently, suspensions of inside-out follicles were used in which the apical plasma mem-branes of follicle cells were directed toward the culture medium and separated by tight junctions from the lateral cell surfaces (Herzog and Miller, 1981). The tightness of the epithelial monolayer and the stable inside-out polarity are prereq-uisites to study endocytosis selectively from the apical plasma membrane. It is the purpose of this article to describe the pathways of endocytosis in thyroid follicle cells with particular reference to results obtained from work with inside-out follicles.

II. The Polar Organization of Follicle Cells (Table I)

The wall of thyroid follicles is composed of epithelial cells which form a tight monolayer and which separate the follicle lumen from the extrafollicular space. The polar organization of follicle cells is recognized (1) functionally, by the bidirectional transport in the cell of thyroglobulin to and from the apical cell surface (for schematic representation, see Fig. 26), and (2) structurally, by the asymmetric distribution of organelles and of cell surface components in the apical and the basolateral cell surfaces. This membrane asymmetry correlates with the presence of tight junctions which have been shown in other epithelia to act as a barrier to lateral diffusion of membrane constituents (Dragsten *et al.*, 1981).

The apical plasma membrane is characterized by the presence of thyroperox-idase (Tice and Wollman, 1974), of aminopeptidase (Hovsépian *et al.*, 1982), and by its function in the iodination of thyroglobulin (Ekholm and Wollman, 1975; Ekholm, 1981). It is the most heterogeneous area of the surface membrane of follicle cells being subdivided into several microdomains which differ in structure and composition. Numerous irregularly arrayed microvilli increase the surface area of the apical plasma membrane ~4- to 5-fold (Wollman and Loewenstein, 1973; Ericson and Engström, 1978). Microvilli carry a prominent glycocalyx (Herzog and Miller, 1981); frequently, coated pits are located at the bases of microvilli. Coated pits are not restricted to the apical plasma membrane but are observed on the basolateral cell surfaces as well. Pseudopods formed by follicle cells after stimulation with TSH. They are located on the peripheral portion of the apical plasma membrane, often close to the tight junction, and project toward the follicle lumen. The plasma membrane of pseudopods differs compositionally from the other portions of the apical cell surface by the lack of the prominent glycocalyx, of thyroperoxidase (Tice and Wollman, 1974), of NADPH-oxidase (Björkman *et al.*, 1981), and of binding sites for cationized ferritin (see Fig. 9). One to two cilia are located in the center of the apical surface of each follicle cell (Herzog and Miller, 1979a).

The lateral cell surface is separated from the apical plasma membrane by tight junctions and characterized by interdigitations with neighboring cells, and by the presence of desmosomes (Farquhar and Palade, 1963; Bernstein and Wollman, 1976) and of gap junctions (Tice *et al.*, 1976; Thiele and Reale, 1976).

The basal plasma membrane is associated with the basement membrane. Both the lateral and the basal cell surfaces contain Na^+, K^+-ATPase (Wolff, 1964) and operate in the active uptake of iodide (Chambard *et al.*, 1983).

III. Inside-Out Follicles as a Model System for Studies on Endocytosis

A. FORMATION OF INSIDE-OUT FOLLICLES

Thyroid tissue of various species can be dissociated into large follicle seg-ments (Figs. 1 and 2) by digestion with collagenase and the application of shearing forces (Herzog and Miller, 1979a; Denef *et al.*, 1980). When kept in suspension, the follicle segments from pig thyroid gland form closed follicular structures (Herzog and Miller, 1979a). Within 24 hours of suspension culture, all follicle segments are transformed into follicles (Figs. 1 and 3–5) whose polarity is reversed with the apical plasma membranes of follicle cells directed toward the culture medium and the basal cell surfaces limiting the newly formed central

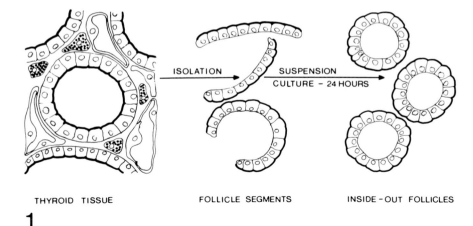

THYROID TISSUE   FOLLICLE SEGMENTS   INSIDE-OUT FOLLICLES

1

FIG. 1.   Diagram showing the formation of inside-out follicles from pig thyroid gland. Thyroid tissue is dissociated into segments of the follicle wall. Within 24 hours of suspension culture, follicle segments bend and form closed follicular structures which all exhibit a reversed polar organization. The apical plasma membranes of follicle cells are accentuated by dark lines to denote the reversed polar organization of inside-out follicles as compared to follicles in thyroid tissue.

cavity (Figs. 1 and 4 and 5). Such inside-out follicles are tight against the extracellular exchange of macromolecules between the culture medium and the central cavity; they can be stimulated with thyrotropin (TSH) and they remain structurally and functionally polarized and single-walled for several days after follicle closure and during handling (e.g., pipetting or exposure to various tracers) (Herzog and Miller, 1981). Hence, inside-out follicles appear much more stable than rightside-out follicles which can be isolated in an intact state (Herzog and Miller, 1979a; Denef *et al.*, 1980; Nitsch and Wollman, 1980), but are extremely fragile: they easily rupture, release their luminal content, and ultimately form inside-out follicles. The inside-out polarity appears, therefore, under normal culture conditions the most stable conformation of thyroid follicles *in vitro*.

It is unknown at present why the epithelial monolayer of thyroid segments always bends in such a direction that gives rise to the formation of inside-out follicles. However, there are two factors which may support the inversion.

1. The apical plasma membrane is particularly rich in binding sites for polycationic markers such as DEAE-dextran (Herzog and Miller, 1979b) indicating that the concentrations of anionic surface charges are higher on the apical cell surface than on the other plasma membrane domains (Table I). It is assumed that the repulsion of negative surface charges can force the epithelial sheet to bend in one direction.

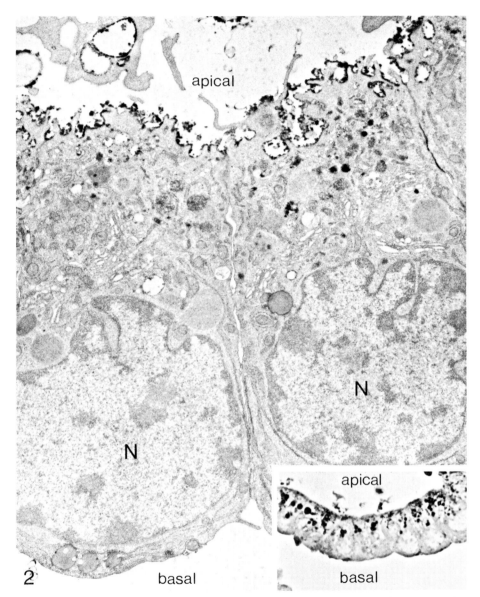

Fig. 2.   Portions of follicle segments stimulated with thyrotropin and incubated for 10 minutes in culture medium containing horseradish peroxidase. In such preparations, all cell surface domains are accessible by the tracer, but the apical plasma membrane predominates in endocytosis as indicated by the presence of the tracer within vesicles underneath the apical cell surface. This points to the functional heterogeneity of surface domains in thyroid follicle cells. N, Nucleus. ×9700; inset, ×1100.

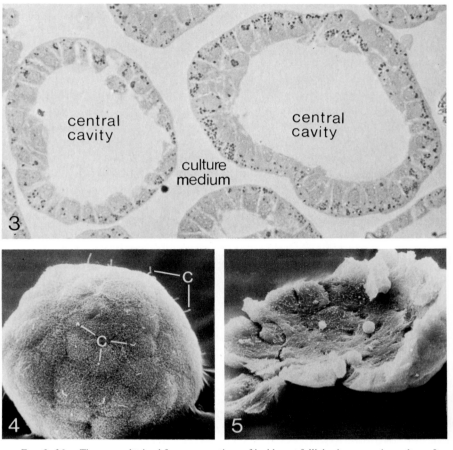

Fɪɢ. 3–26.   These are obtained from preparations of inside-out follicles in suspension culture. In such preparations, only the apical plasma membranes are accessible to tracers.

Fɪɢ. 3.   Inside-out follicles in suspension culture. The apical plasma membrane is directed toward the culture medium and characterized by the preferential location of lysosomes in the apical cytoplasm. The basal cell surfaces border the newly formed central cavity. ×350.

Fɪɢ. 4 ᴀɴᴅ 5.   Scanning electron micrographs from inside-out follicles. The apical plasma membrane is identifiable by the presence of cilia (c) and by the microvilli projecting toward the outside (Fig. 4). When follicles are opened after critical point drying and coated with gold (Fig. 5) the basal cell surfaces can be viewed which lack the characteristic microvilli and the cilia. 4, ×1200; 5, ×650.

TABLE I

POLAR DISTRIBUTION OF CELL SURFACE COMPONENTS IN FOLLICLE CELLS[a]

Cell surface area	Topographical characteristics	Structural characteristics	Compositional characteristics
Apical	Directed toward the thyroglobulin-containing lumen	1–2 centrally located cilia; microvilli[b] with prominent glycocalyx[c]; coated pits[d]; pseudopods[e]	Thyroperoxidase[d]; NADPH oxidase[d]; aminopeptidase[e]; excess of anionic sites[f]
Lateral	Border to the intercellular cleft	Gap junctions[g]; desmosomes[h]; fingerlike projections which interdigitate with neighboring cells; coated pits	Location of Na$^+$,K$^+$-ATPase and transport system for iodide[i]; binding sites for fibronectin[j]
Basal	Attachment to the basement membrane	Fingerlike projections, coated pits	

[a]The space between apical and lateral marks the location of tight junctions which separate the follicle lumen from the extrafollicular space and the apical from the basolateral plasma membranes.

[b]Wollman and Loewenstein (1973) and Ericson and Engström (1978).

[c]Herzog and Miller (1979a, 1981).

[d]Nadler et al. (1962); Wollman et al. (1964). The surface membrane of pseudopods differs from the main part of the apical cell surface by the lack of thyroperoxidase (Tice and Wollman, 1974), of NADPH oxidase (Björkman et al., 1981), of binding sites for cationized ferritin and of a prominent glycocalyx (see also Fig. 9). Pseudopods are formed when follicles are stimulated with TSH.

[e]Hovsépian et al., (1982).

[f]Herzog and Miller (1979b).

[g]Thiele and Reale (1976) and Tice et al. (1976).

[h]Farquhar and Palade (1963) and Bernstein and Wollman (1976).

[i]The transport of iodide into follicle cells is functionally coupled to the Na$^+$,K$^+$-ATPase (Wolff, 1964) and occurs only at the basolateral plasma membrane (Chambard et al., 1983).

[j]Giraud et al. (1981).

2. The contractive forces of the terminal web which is localized underneath the apical plasma membrane (Gabrion et al., 1980; Herzog and Miller, 1981) and which is in connection with a circumferential bundle of microfilaments (Bernstein and Wollman, 1976) could force the cells in the periphery of an epithelial sheet to bend in such a way that the apical plasma membrane forms a convexity. Contraction of the terminal web, perpendicular to the microvilli, has been shown to bend the edges of epithelial cells (Burgess, 1982) and of isolated brush border preparations (Rodewald et al., 1976; Hirokawa et al., 1983).

The inside-out polarity of follicles in vitro correlates with a reversed vectorial transport since thyroglobulin does not accumulate in the central cavity but is released into the culture medium (Herzog and Miller, 1981). There it becomes detectable after incubation of inside-out follicles with L-[^{3}H]leucine (see Fig. 6), D-[^{3}H]galactose, or ^{125}I (Herzog, 1983b). The sequential incorporation of leucine in the rough endoplasmic reticulum, of galactose in the Golgi complex (Whur *et al.*, 1969), and of iodine at the apical plasma membrane (Ekholm and Wollman, 1975) allows the conclusion that a reversed functional polarity is established.

The reversed functional polarity is also observed when thyroglobulin is added to suspensions of inside-out follicles. Thyroglobulin is internalized from the culture medium by the apical plasma membrane and transferred to lysosomes. This results in the release of thyroxine and triiodo thyronine into the culture medium at increased rates when follicles are stimulated with thyrotropin (Fig. 7) (Herzog and Miller, 1981).

B. APPLICATION OF TRACERS

The labeling of the apical plasma membrane of inside-out follicles requires simply the addition of the tracer to the culture medium. However, certain precautions must be taken in the use of membrane markers (for general discussion see Herzog and Farquhar, 1983). Cationized ferritin binds electrostatically to sialic acid and to other anionic residues on the cell surface (Danon *et al.*, 1972) and is used as a nonspecific membrane marker. Cationized ferritin may also form aggregates with anionic proteins in the culture medium such as fetal calf serum or newly released thyroglobulin and lose its affinity to binding sites on the plasma membrane. To achieve satisfactory binding to the apical cell surface, inside-out follicles have to be washed extensively in PBS prior to the application of the tracer.

If applied at concentrations of 10 to 50 μg/ml, cationized ferritin does not disturb the architecture of inside-out follicles as reported for other epithelial systems (Quinton and Philpott, 1973) and is readily internalized by follicle cells. To achieve comparable uptake of native ferritin which is anionic and which does not bind to the apical plasma membrane, ~1000-fold higher concentrations are required, i.e., 10–50 mg/ml (Herzog and Farquhar, 1983).

Horseradish peroxidase binds to the plasma membrane (Fig. 2). The binding is specifically inhibited by 10^{-3} M mannose and may indicate, therefore, the presence of a mannose-receptor (Sly and Stahl, 1978; Straus, 1983) in thyroid follicle cells.

Thyroglobulin can be used as a tracer when coupled to an electron-opaque marker such as colloidal gold with a particle diameter of ~20 nm (see Fig. 18) or when biosynthetically labeled with L-[^{3}H]leucine (see Fig. 19) or with ^{125}I (Herzog, 1983b). The mildest effects on the structure of thyroglobulin are probably achieved with radiolabeling. This, however, requires autoradiographic detec-

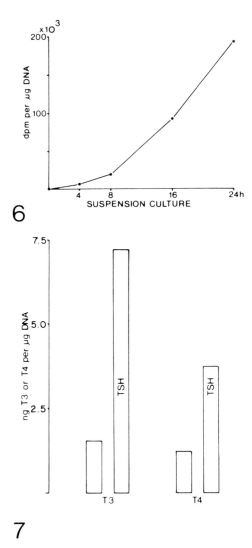

FIG. 6.   In the presence of [³H]leucine, [³H]thyroglobulin is synthesized and released by inside-out follicles into the culture medium (compare with Fig. 19).

FIG. 7.   Inside-out follicles (suspended in culture medium containing 20 mg/ml thyroglobulin) internalize thyroglobulin and release triiodothyronine (T3) and thyroxin (T4). This process is stimu-lated by TSH.

tion which has the drawback to be of limited spatial resolution. A much higher resolution is gained by the use of the thyroglobulin–gold complex. It cannot be excluded, however, that the structure of the thyroglobulin molecule is changed after its attachment to gold particles. Therefore, different labeling procedures are required to follow the endocytic pathways of thyroglobulin.

IV. Formation of Endocytic Vesicles at the Apical Plasma Membrane

Endocytosis occurs on all cell surfaces of thyroid follicle cells; however, the apical plasma membrane is far more active in endocytosis than the other membrane areas. This is best observed in open follicle segments in which all plasma membrane domains of follicle cells can be reached and is exemplified with horseradish peroxidase as tracer: Upon stimulation with thyrotropin, horseradish peroxidase is rapidly internalized on the apical cell surface whereas the other membrane domains are less active in endocytosis (Fig. 2).

Two different mechanisms for the endocytic uptake from the apical surface of follicle cells are under discussion: (1) macropinocytosis by pseudopods and (2) micropinocytosis by coated pits at the bases of microvilli. Pseudopods usually carry one or several vacuoles containing internalized tracer. However, details on the uptake-mechanism by pseudopods are unknown (for discussion on the possible function of pseudopods in endocytosis see Wollman, 1969, and Ericson, 1981). Furthermore, micropinocytosis by coated pits appears to be the major mode of entry of thyroglobulin (Seljelid et al., 1970). The following discussion concentrates, therefore, on micropinocytosis at the apical plasma membrane.

Appropriate ligands such as cationized ferritin, wheat germ agglutinin, or antibodies to aminopeptidase are, at 4°C, evenly distributed on the membranes of microvilli (Fig. 8) and of coated pits. Soon after raising the temperature to 37°C, the ligands undergo a characteristic redistribution on the apical plasma membrane. For example, cationized ferritin collects mainly in or close to coated pits at the bases of microvilli (Figs. 9 and 16). This process is temperature dependent but does not require stimulation with TSH.

Ten minutes after the addition of TSH to suspensions of inside-out follicles,

FIGS. 8 AND 9. Apical cell surface of inside-out follicles labeled with cationized ferritin (CF). CF particles are evenly distributed at 4°C or after fixation of follicles with glutaraldehyde (Fig. 8) but concentrated in or close to coated pits at the bases of microvilli when follicles are warmed to 37°C (Fig. 9). Pseudopods (PP) are located close to the tight junction (ZO); their surface membrane differs from that of microvilli (mv) as it lacks a prominent glycocalyx and binding sites for CF (see also Table I). Upon stimulation with thyrotropin, coated pits (inset, cp) detach and form coated vesicles (cv). Apparently, coated pits shed their coat rapidly because CF particles are observed in smooth surfaced vesicles after 1 minute of endocytosis. 8, ×70,000; 9, ×29,000; inset 9, ×74,000.

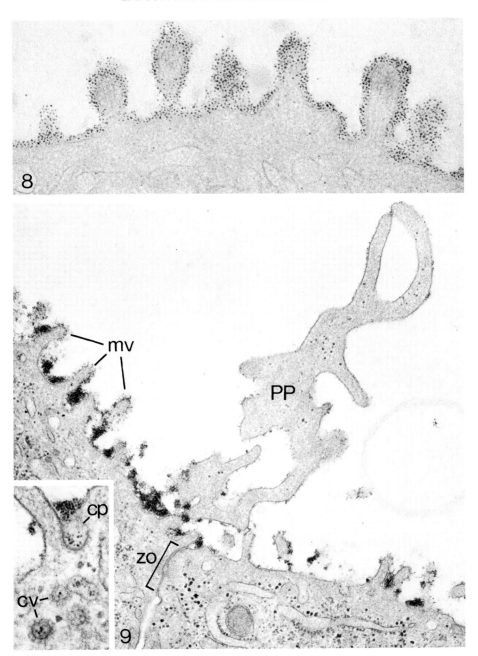

coated pits detach (Herzog and Miller, 1979a, 1981) to form coated vesicles (inset Fig. 9), which appear to shed their coat rapidly; after ~1 minute of endocytosis smooth-surfaced endocytic vesicles carrying the tracer are observed. Pastan and Willingham (1981) reported the appearance of α_2-macroglobulin in uncoated vesicles within 20 seconds of endocytosis.

The internalization of membrane constituents is a highly controlled event and appears to exclude resident proteins which remain in the plasma membrane and which are segregated during endocytosis from migrant proteins (Brown et al., 1983). In thyroid follicle cells, it has been shown that thyroperoxidase activity is absent from the membranes of endocytic vesicles and of endosomes (Fig. 14). Thyroperoxidase, an integral membrane constituent (Neary et al., 1978) and present in the apical cell surface (Tice and Wollman, 1974), may be excluded from endocytosis (Herzog and Miller, 1980) and represents, therefore, a resident membrane protein. However, it is also possible that thyroperoxidase is internalized in an inactive state. At any rate, the absence of thyroperoxidase activity from endocytic compartments is of considerable functional significance because it indicates that the peroxidase-dependent iodination of thyroglobulin does not operate in endocytic compartments.

Thyroglobulin binds to the apical plasma membrane as shown in Figs. 20 and 23. This binding supports the idea that receptors are present on the surface of follicle cells which recognize asialothyroglobulin (Consiglio et al., 1979). In addition, preferential binding and uptake of highly iodinated thyroglobulin species as observed by van den Hove et al. (1982) suggest a certain selectivity during endocytosis of thyroglobulin. The presence of binding sites does not necessarily imply that thyroglobulin is internalized only by absorptive endocytosis. The high concentration of thyroglobulin in follicle lumena of the intact gland with mean values at 110 mg/ml (Smeds, 1972) and highest concentrations up to 400 mg/ml (Hayden et al., 1970) suggests the occurrence of additional uptake by fluid phase pinocytosis. It is, therefore, proposed that endocytic vesicles are able to carry thyroglobulin by two distinct modes: (1) thryoglobulin absorbed to the inner surface of the vesicle membrane and (2) thyroglobulin molecules suspended in the vesicle lumen.

V. Pathway to Endosomes and Lysosomes

The main pathway of endocytosis in thyroid follicle cells is directed toward the lysosomes where thyroid hormones are released by proteolysis of thyroglobulin. The route to lysosomes begins with the formation of endosomes which are presumed to be transformed into lysosomes.

A. ENDOSOMES AND COLLOID DROPLETS

Endosomes are prelysosomal compartments (Korn, 1975; Helenius *et al.*, 1980, 1983); they are characterized by an acidic (Tycko and Maxfield, 1982) and electron-lucent interior and delimited by a membrane which is often connected with tubular elements (Hopkins and Trowbridge, 1983). Endosomes are particularly rich in binding sites or receptors derived from the plasma membrane and, therefore, also termed "receptosomes" (Pastan and Willingham, 1983). The term receptosome differs conceptually from that of the endosome by the assumption that coated pits are permanently connected to the cell surface and that ligands are carried into the cell by vesicles (receptosomes) which bud off from the necks of coated pits (Pastan and Willingham, 1983). Other terms synonymous to endosomes are "phagosome" (Straus, 1964), "resorptive vacuole" (Miller and Palade, 1964), "pinosome," and "endocytic vacuole" (Steinman *et al.*, 1983).

Endosomes in cells of inside-out follicles are $\sim$1 μm in diameter, located in the apical cytoplasm (see Fig. 16), and usually surrounded by several endocytic vesicles. Endosomes are reached by endocytic vesicles within 2–5 minutes of endocytosis. After 1 hour of endocytosis, $\sim$15% of all cationized ferritin internalized are found in endosomes. Ligands such as cationized ferritin and thyroglobulin remain attached to the inner surface of the endosomal membrane (Figs. 10–14, 16, and 24). Thyroperoxidase activity is demonstrable in compartments of the export pathway (Strum and Karnovsky, 1970; Novikoff *et al.*, 1974), i.e., the rough endoplasmic reticulum (Fig. 14), some Golgi-cisternae, secretion granules, and part of the apical plasma membrane (Tice and Wollman, 1974). It appears absent, however, from endocytic vesicles, endosomes (Fig. 14), and lysosomes. The reactivity of thyroperoxidase allows, therefore, distinguishment between the compartments of the endocytic pathway and of the export route.

Acid phosphatase and presumably other lysosomal hydrolases are always absent from endosomes (Figs. 12 and 13). Their prelysosomal nature is also shown by the observation that endosomes are reached by the internalized tracer before it is seen in lysosomes. Multivesicular bodies which appear to be formed from endosomes (Holtzman, 1976; Haigler *et al.*, 1979) are often observed in the apical portion of thyroid follicle cells. As reported for other cell types, e.g., epithelial cells of the ductus deferens (Friend and Farquhar, 1967), multivesicular bodies in thyroid follicle cells also contain cytochemically detectable acid phosphatase and represent, therefore, lysosomes. However, acid phosphatase-negative multivesicular bodies (multivesicular endosomes) have been reported to occur in reticulocytes and to constitute there a recycling compartment because of their ability to fuse again with the plasma membrane (Harding *et al.*,

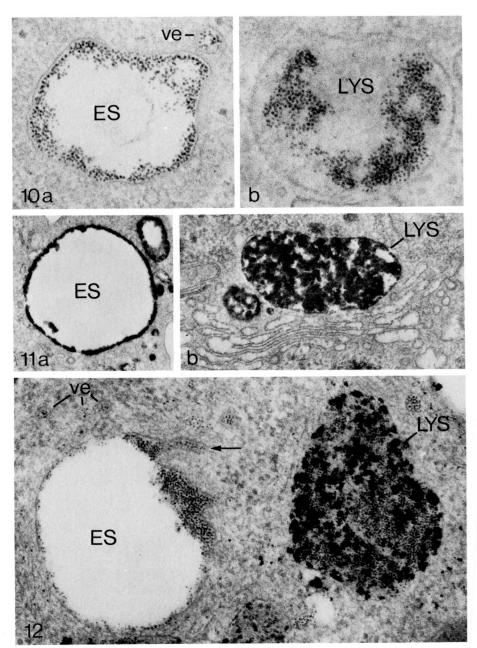

FIGS. 10–14. Characteristics of endosomes in thyroid follicle cells.

FIGS. 10–13. Endosomes are reached at 2–5 minutes of endocytosis and characterized by the adherence of the membrane marker (CF in Figs. 10a, 12, and 13; horseradish peroxdiase attached to a mannose-receptor in Fig. 11a) to the inner surface of the endosomal membrane, by the absence of

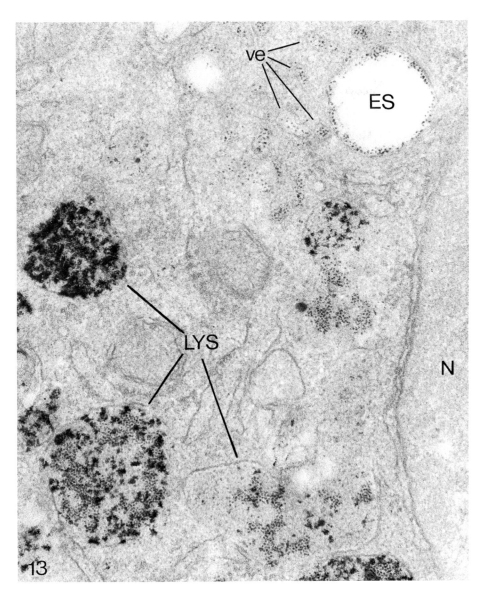

acid phosphatase (Figs. 12 and 13), and by tubular extensions (arrow in Fig. 12). Usually, endosomes (ES) are surrounded by numerous endocytic vesicles (ve in Figs. 10a, 12, and 13) which are free of acid phosphatase reaction product (Figs. 12 and 13). In lysosomes (LYS), the tracers dissociate from the membrane and accumulate in the lysosomal matrix (Figs. 10b, 11b, 12, and 13) which reacts positively for acid phosphatase (Figs. 12 and 13). Note that the amount of both the internalized tracer and reaction product of acid phosphatase can vary in different lysosomes of the same cell (LYS in Fig. 13). 10a, ×90,000; 10b, ×100,000; 11a, ×29,000; 11b, ×43,000; 12, ×72,000; 13, ×80,000.

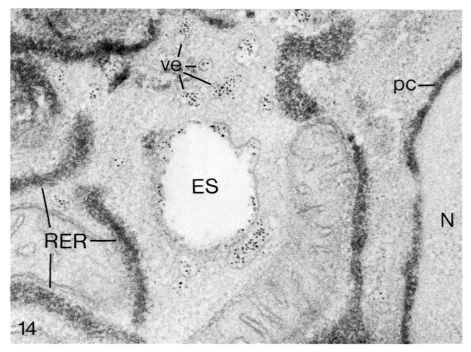

FIG. 14.   Thyroperoxidase activity is absent from endocytic vesicles (ve) and from endosomes (ES) and observed only in compartments of the export route for thyroglobulin, e.g., the rough endoplasmic reticulum (RER) including the perinuclear cisterna (pc). N, Nucleus. ×84,000.

1983). In thyroid follicle cells, acid phosphatase-negative multivesicular bodies and their fusion with the plasma membrane have not been observed and other mechanisms of membrane recycling appear to operate (see Sections VI and VII,A).

In cells from intact tissue where the apical cell surface is exposed to thyroglobulin at concentrations up to 400 mg/ml (Hayden *et al.*, 1970; Smeds, 1972), endosomes with the electron-lucent interior characteristic for follicle cells *in vitro* are not observed. Instead, within a few minutes after stimulation with thyrotropin numerous spherical organelles (1–2 μm in diameter) with a fine granular content are formed which are usually referred to as colloid droplets (Wollman *et al.*, 1964; Wollman, 1969). Radiolabeled thyroglobulin becomes detectable in colloid droplets 5–10 minutes after endocytosis (Wollman and Spicer, 1963; Stein and Gross, 1964; Seljelid *et al.*, 1970) at concentrations which correspond to those observed in the follicle lumen (Bauer and Meyer, 1964). Microinjection studies by Seljelid *et al.* (1970) have shown that colloid droplets are formed primarily by micropinocytosis. Colloid droplets have not yet acquired acid phosphatase activity (Wetzel *et al.*, 1965) and do not contain cytochemically demonstrable thyroperoxidase (Tice and Wollman, 1974). They constitute a prelysosomal compartment and correspond, therefore, to the endosomes described in follicle cells *in vitro*.

B. LYSOSOMES

Lysosomes in thyroid follicle cells *in situ* are characterized by the presence of both ingested thyroglobulin and cytochemically detectable acid hydrolases. Colloid droplets acquire acid phosphatase within 15 minutes after thyrotropin-induced endocytosis by fusion with primary lysosomes (Wollman *et al.*, 1964; Wetzel *et al.*, 1965). Newly formed secondary lysosomes are usually large and located in the apical portion of follicle cells. This stage is followed by shrinkage due to degradation of thyroglobulin and by an increase in density resulting in the formation of dense granules, which are preferentially located at the basal portions of follicle cells. It has been suggested that dense granules have a "cyclic existence" in that stimulation with thyrotropin results in their movement back to the apical portion of follicle cells where they fuse with colloid droplets (endosomes) containing newly internalized thyroglobulin (Wollman, 1969).

In cells of inside-out thyroid follicles, a similar distribution with small lysosomes at the cell bases and larger lysosomes in the apical cytoplasm has been observed. Direct evidence for the cycling of lysosomes as suggested by Wollman (1969) is not available. The transformation of endosomes to lysosomes is indicated by the acquisition of acid phosphatase activity (Figs. 12 and 13) and by the accumulation of membrane markers in the lysosomal matrix (Figs. 10–13, 15, and 16). Apparently, membrane markers which are bound to the inner surface of the endosomal matrix detach from the membrane when lysosomal enzymes appear due to hydrolysis of the ligand.

Lysosomes in the cells of inside-out follicles are reached by the tracers at ~5 minutes after TSH-stimulated endocytosis. This suggests a life-time for endosomes of only a few minutes, if all endosomes are transient organelles. Alternatively, it has been suggested that endosomes are stable structures which receive internalized material and deliver it by a vesicular mechanism to lysosomes (for detailed discussion see Helenius *et al.*, 1983). With cationized ferritin as a tracer, ~50% of all tracer particles internalized are observed in lysosomes after 1 hour of endocytosis.

The fate of membrane retrieved during endocytosis is still unclear because ligands detach from the lysosomal membrane thereby losing the ability to function as a membrane marker. It has been suggested, however, that a vesicular recovery route of internalized membrane from lysosomes to the stacked Golgi-cisternae exists which are reached at ~30 minutes (Herzog and Miller, 1979a, 1981; see also Section VI).

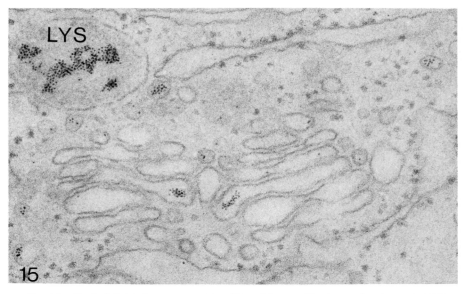

FIG. 15.   Stacked Golgi-cisternae containing CF particles at ~30 minutes of endocytosis. LYS, Lysosomes. ×71,000.

It is unknown, at present, by which mechanism the thyroid hormones are released from the lysosomes and from the follicle cells. A vesicular shuttle between lysosomes and the basolateral plasma membrane would allow the simultaneous release of thyroid hormones and of thyroglobulin (see transepithelial transport, Section VII). Alternatively, the diffusion of iodothyronines from secondary lysosomes into the cytoplasma and from there into the extracellular space has been proposed (Wollman, 1969).

VI. The Golgi Complex as an Organelle Where Export and Endocytic Pathways Meet

The central feature of the Golgi complex is a stack of 4–7 closely apposed and flattened cisternae. Functionally, the Golgi complex is considered an obligatory station on the export pathway of secretory proteins and of newly formed plasma

FIG. 16.   Apical portions of epithelial cells from a thyrotropin-stimulated inside-out follicle after 30 minutes of endocytosis. Endocytic vesicles formed at the apical plasma membrane (PM) have transferred CF particles to endosomes (ES), lysosomes (LYS), the stacked Golgi-cisternae (arrowhead, G), and the lateral cell surfaces (arrows). The transcytotic vesicles bypass the tight junctions (ZO) and acquire a coat on their cytoplasmic surface after their fusion with the lateral (inset, cp) or basal plasma membranes. N, Nucleus. ×30,000; inset, ×54,000.

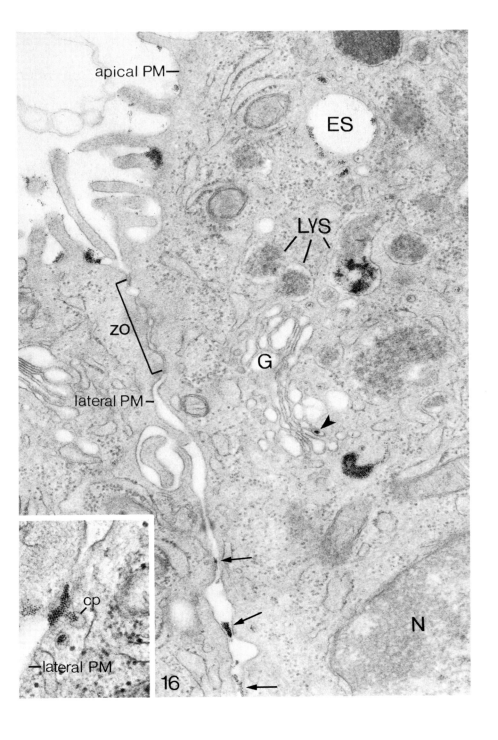

apical PM—

ES

LYS

ZO

lateral PM—

G

cp

—lateral PM

16

N

membrane. Topographically, it is interposed between the rough endoplasmic reticulum on the cis- or entry-face and the secretion granules on the trans- or exit-face (Tartakoff, 1980; Herzog, 1980; Farquhar and Palade, 1981). In thyroid follicle cells, the Golgi complex functions in the packaging and in the terminal glycosylation of thyroglobulin. Galactose (Whur *et al.*, 1969) and fucose (Haddad *et al.*, 1971) are added in the Golgi complex to the carbohydrate side chain of thyroglobulin (for review see Leblond and Bennett, 1977). Although thyroperoxidase is present in the membranes of one or two cisternae on the trans-side of the Golgi complex (Strum and Karnovsky, 1970; Novikoff *et al.*, 1974; Herzog and Miller, 1979a), iodination which is one of the most important post-translational modifications of thyroglobulin does not take place before it reaches the apical plasma membrane (Ekholm and Wollman, 1975; Ekholm, 1981).

In recent years it became clear that the Golgi complex is not only a station where the exportable products and membrane constituents are modified while in transit, but also an organelle which is reached directly or indirectly by endocytic vesicles and is composed, therefore, in part of membrane components derived from the cell surface (Herzog and Farquhar, 1977; Herzog, 1980; Farquhar, 1981). The participation of the Golgi complex in endocytosis has been interpreted in other secretory cells (Herzog and Farquhar, 1977; Farquhar, 1978; Herzog and Reggio, 1980; Ottosen *et al.*, 1980) as an indication of reutilization of membrane during the secretory process.

In thyroid follicle cells, studies on endocytosis have shown that the main endocytic pathway is directed to colloid droplets (endosomes) and to lysosomes (see Section V). Some tracers used previously in other cells for studies on endocytosis and membrane recycling (Herzog and Ferquhar, 1983) have also reached the stacked Golgi-cisternae of thyroid follicle cells. Cationized ferritin applied to the apical cell surface appears ~30 minutes after the thyrotropin-induced endocytic uptake in the stacked Golgi-cisternae (Fig. 15) where the tracer is preferentially but not exclusively located in the peripheral rims (Herzog and Miller, 1979a, 1981). Because endosomes and lysosomes are reached by the tracer earlier than the Golgi-cisternae, it has been speculated that some cationized ferritin traces a pathway from endosomes or lysosomes to the stacked Golgi-cisternae. Support for this idea is obtained from experiments with complexes of cationized feritin and latex particles (1 μm in diameter). After their ingestion and transfer to endosomes and lysosomes, the complex dissociates; the latex remains in lysosomes whereas cationized ferritin is found later, at ~30 minutes, also in the stacked Golgi-cisternae (Herzog and Miller, 1979a). All cisternae of the Golgi complex may participate in the endocytic uptake. This includes the one or two thyroperoxidase-containing cisternae on the trans-side of the Golgi stacks (Herzog and Miller, 1979a).

The number of tracer particles observed in Golgi cisternae is low and does not exceed 1% of the total amount internalized from the apical plasma membrane.

Most of the membrane portions may have lost their tracer after fusion with the lysosomal membrane and the number of cationized ferritin particles in the Golgi complex may not correspond to the amount of membrane transferred. It is presumed that the presence of tracer particles in stacked Golgi-cisternae does not result from a chance event but documents a pathway by which components of the apical cell surface are returned to a biosynthetic compartment. Plasma membrane constituents could be reglycosylated or modified in the Golgi complex (Farquhar, 1981) before their reutilization in the secretory cycle. This would be of particular importance after passage through the lysosomal compartment.

Thyroglobulin which binds to the apical plasma membrane of follicle cells (Figs. 20 and 23) is also transferred first to endosomes and lysosomes (Herzog, 1983a), but it is never observed in stacked Golgi-cisternae. Apparently, thyroglobulin dissociates completely from the endocytic membrane which is recovered and transferred at least in part (as shown with cationized ferritin as tracer, see above) to Golgi-cisternae. The compartment of uncoupling receptor from ligand (CURL: Geuze et al., 1983) is therefore assumed to be located in endosomes or lysosomes or in an intermediate organelle. A direct recycling pathway from endosomes and lysosomes back to the apical plasm membrane may exist but has not been described in thyroid follicle cells.

VII. Release of Thyroglobulin into the Circulation

Freeze fracture studies of thyroid follicles have shown that the extrafollicular space is sealed toward the thyroglobulin-containing lumen by a continous belt of zonulae occludentes which are composed of a complex network of 5–8 strands (Tice et al., 1975; Thiele and Reale, 1976). Even during mitosis, the tight junctions remain intact between newly formed daughter cells and the neighboring cells and no signs of gaps are observed through which gross leakage of colloid could possibly occur (Zeligs and Wollman, 1977a,b). The presence of tight junctions correlates with the high transepithelial resistance of follicles in the resting thyroid gland (Williams, 1969). It is concluded that the epithelial wall is tight and thyroglobulin is unable to leave the follicle lumen by diffusion. Accordingly, thyroglobulin was considered to be confined to the thyroid gland. However, with the aid of specific radioimmunoassays, it was possible to show the release of undegraded thyroglobulin into the thyroid lymphatic systems (Daniel et al., 1967) and its appearance in the circulation (Roitt and Torrigiani, 1967) of man and several mammalian species (Table II). It was also observed that the thyroglobulin levels can be raised by stimulation with TSH (Uller et al., 1973) and suppressed by the addition of thyroid hormones (van Herle et al., 1973). Hence, it became clear that the functional state of thyroid follicle cells regulates the release of thyroglobulin. The cellular mechanism of this transfer from the lumen through the follicle wall remained largely unexplored.

TABLE II

Serum Thyroglobulin Levels in Rat and Man

	Conditions	Serum thyroglobulin (ng/ml)[a]	References
Rat	Normal[b]	~100	van Herle et al. (1975)
	TSH stimulation	~330	van Herle et al. (1975)
	Experimentally induced goiter[c]	~150	van Herle et al. (1975)
Man	Normal[d]	~5	van Herle et al. (1973); Uller et al. (1973); Pezzino et al. (1978)
	TSH stimulation	~20	Uller et al. (1973, 1977)
	Goiter	~35	Pezzino et al. (1978)
		~380	Uller et al. (1973)
	Hyperthyroidism (Graves' Disease)	~85	Uller et al. (1973)
	Differentiated thyroid carcinoma	~140	van Herle et al. (1975)
	Subacute thyroiditis	~140	van Herle et al. (1973)

[a]Approximate mean values. For detailed information see literature cited.

[b]Values may differ among various strains with mean values of ~80 ng/ml for Fisher rats and ~220 ng/ml for Sprague–Dawley rats (Izumi and Larsen, 1978).

[c]Goiter in rats was induced either by chronic iodine deficiency or by administration of propylthiourazil (van Herle et al., 1975).

[d]Women, particularly in late pregnancy, have higher serum thyroglobulin levels than men (Pezzino et al., 1977; van Herle et al., 1973). There is no age-related variation (Torrigiani et al., 1969).

A. Vesicular Transport Across Thyroid Follicle Cells (Transcytosis)

Experiments with cationized ferritin showed that in cells of inside-out follicles a vesicular transport mechanism (transcytosis) operates from the apical plasma membrane to the basolateral cell surfaces. Cationized ferritin remains attached to the membrane portions inserted into the basal or lateral plasma membranes. Statistically, the amount of tracer particles accumulating on the basolateral cell surface is directly related to the number of transcytotic vesicles inserted there. Hence, the rate of tracer accumulation allows estimation of the transepithelial transport activity in thyroid follicle cells. The observations indicate that in thyrotropin-stimulated inside-out follicles the transepithelial transport is about 9-fold above that in resting cells (Fig. 17). The lateral plasma membrane is always reached earlier (at ~11 minutes) and by more vesicles than the basal cell surface into which transcytotic vesicles become inserted at ~16 minutes after endocytosis. Transcytosis is temperature dependent and stopped at ~15°C and below (Herzog, 1983a). After 1 hour of endocytosis, ~10% of all cationized ferritin internalized is transferred across the epithelial wall.

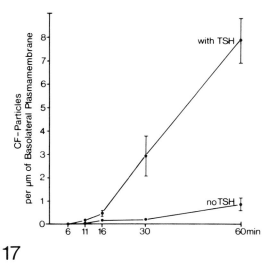

17

FIG. 17. Effect of TSH on the transepithelial vesicular transport. Transcytosis occurs in the resting follicle (no TSH) but is increased about 9-fold when follicles are stimulated (with TSH). Results are obtained by morphometric analysis of CF particles translocated across the follicle wall and given as mean values from nine follicle cells ± SD (for details see Herzog, 1983a).

The transcytotic vesicles are also able to carry thyroglobulin across the follicle wall as shown using thyroglobulin as a tracer. For this purpose, thyroglobulin was either attached to gold particles (20 nm in diameter) (Fig. 18) or radiolabeled biosynthetically by a separate follicle preparation in the presence of ^{125}I or [3H]leucine (Fig. 19). In both cases, thyroglobulin is absorbed to the apical cell surface (Figs. 20 and 23), internalized, and transported mainly to endosomes and lysosomes where it is observed begining 5 minutes after endocytosis (Figs. 21 and 24). Internalized thyroglobulin is never detected in Golgi-cisternae but small portions are released at the basolateral cell surfaces (Figs. 22 and 25) where they detach from the membrane and accumulate in the central cavity.

The transcytotic vesicles appear smooth surfaced whereas endocytic pits at the apical (Fig. 9) and exocytic pits at the basolateral (Figs. 16 and 22) membranes are coated on their cytoplasmic surfaces. Hence, the coat appears to be shed during the formation of transcytotic vesicles and not to be regained before insertion into the basolateral cell surfaces.

It is unknown at present whether the transcytotic vesicles migrate directly from the apical to the basolateral plasma membranes or whether they fuse intermittently with other compartments. Intermittent fusion with lysosomes is unlikely because the transcellular transported [3H]thyroglobulin does not differ in molecular weight from the [3H]thyroglobulin before endocytosis, indicating that thyroglobulin is not exposed to proteolysis during transcytosis (Herzog, 1983a).

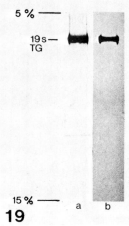

FIGS. 18 AND 19.   Thyroglobulin (Fig. 18a) is used as a tracer after complex formation with 200-Å gold particles (Fig. 18b) or when radiolabeled (biosynthetically in the presence of [³H]leucine). Note that [³H]thyroglobulin (Fig. 19b, right lane) released by inside-out follicles has the same mobility in SDS-gel electrophoresis as thyroglobulin freshly prepared from pig thyroid gland (Fig. 19a, left lane). For tracer studies, [³H]thyroglobulin was extensively dialyzed to remove unbound [³H]leucine. 18a, ×150,000; 18b, ×235,000.

However, transcytotic vesicles could fuse intermittently with endosomes, detach again, and continue their pathway to become inserted in the basolateral cell surface. This indirect pathway is assumed to exist in the small intestine of the newborn rat (Abrahamson and Rodewald, 1981).

FIGS. 20–22.   Transcytosis of thyroglobulin–gold. The thyroglobulin–gold complex (TG-Au) attaches to the apical plasma membrane (PM) and is observed mainly at the bases of microvilli (Fig. 20). Upon stimulation with thyrotropin, TG-Au is internalized and transferred mainly to lysosomes (LYS) but never observed in stacked Golgi-cisternae (G, Fig. 21). A small fraction of TG-Au is also released by the fusion of transcytotic vesicles with the lateral or the basal plasma membrane thereby forming coated pits (cp, Fig. 22). 20, ×49,000; 21, ×31,000; 22, ×35,000.

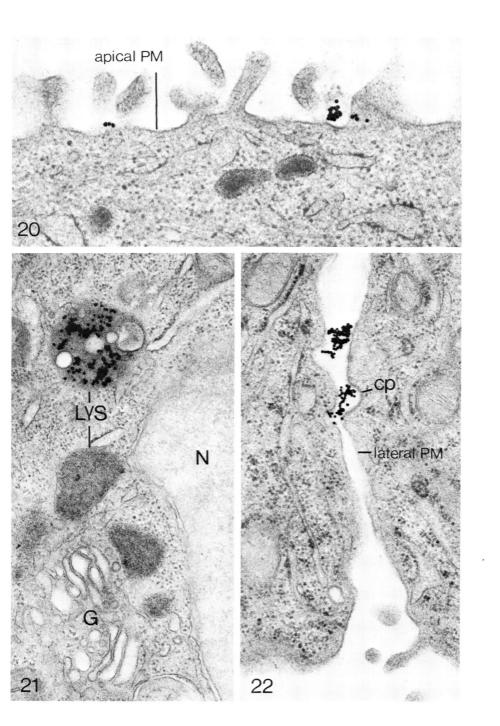

20 apical PM

21 LYS N G

22 cp lateral PM

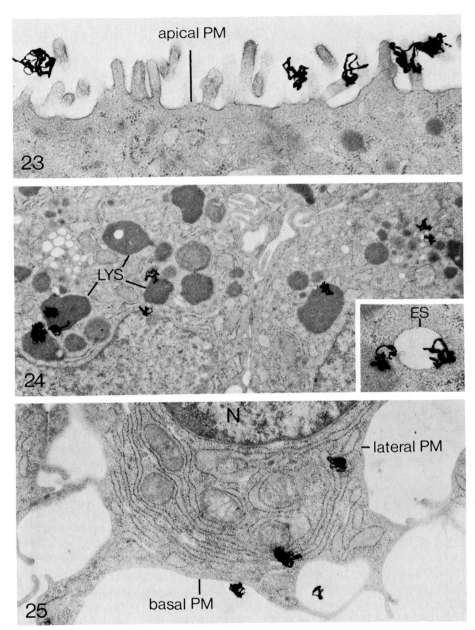

FIGS. 23–25. Transcytosis of [³H]thyroglobulin, autoradiographic analysis. [³H]Thyroglobulin attaches to the apical plasma membrane (PM, Fig. 23). In follicles stimulated with thyrotropin, thyroglobulin is rapidly transferred to endosomes (ES, inset Fig. 24) and to lysosomes (LYS, Fig. 24). A fraction of [³H]thyroglobulin is also released at the basolateral cell surfaces (Fig. 25). N, Nucleus. 23, ×24,000; 24, ×9200; 25, ×14,000.

TABLE III

TRANSEPITHELIAL VESICULAR TRANSPORT IN VARIOUS CELL TYPES

Cell type	Protein transferred	Transfer studied with	References
Epithelial cells of the small intestine	IgG	IgG-ferritin IgG-HRP	Rodewald (1980)
Hepatocytes	IgA	IgA-HRP [^{125}I]IgA	Renston et al. (1980)
Epithelial cells of the choroid plexus	Solutes in the cerebrospinal fluid	Cationized ferritin	van Deurs et al. (1981)
Yolk sac endoderm cells	IgG	IgG-gold	Huxham and Beck (1981)
Thyroid follicle cells	Thyroglobulin (TG)	Cationized ferritin [^{125}I]TG [^{3}H]TG	Herzog (1983a)

Transcytosis has been described in numerous other epithelial cells. Some examples are given in Table III. In all cases, it is assumed that biologically significant proteins are carried across an epithelial monolayer by vesicles which by-pass the tight junctions and the lysosomal compartment. The transepithelial pathway implies the translocation of membrane portions from the apical into the basolateral cell surfaces. Despite this transfer of membrane, the cell surface components in thyroid follicle cells retain their characteristic polar distribution even during stimulation with thyrotropin. Hence, transcytosis must involve efficient sorting of membrane components to preserve follicle cell polarity. Most likely, thyroperoxidase is excluded from endocytosis because it is not detectable in endocytic vesicles (see Section IV; Fig. 14). It has been postulated but remains to be shown that other membrane components enter the transcytotic vesicles and that they are ultimately returned to the same membrane area from which they originate. A vesicular transport from the basolateral to the apical cell surface has been observed (Denef and Ekholm, 1980), but it is unknown whether this route compensates for the transcytotic membrane flow.

Another mechanism for maintenance of follicle cell polarity appears to be associated with the tight junctions located at the boundary between the apical and the basolateral plasma membranes (Table I). Pisam and Ripoche (1976) have reported the redistribution of apical markers over the entire cell surface in dissociated toad bladder epithelial cells. Similarly, the brush border enzymes leucine aminopeptidase and alkaline phosphatase become distributed over the entire cell surface when intestinal cells are isolated (Ziomek et al., 1980). Ca^{2+} is important for maintenance of zonulae occludentes and, therefore, of cell polarity because Ca^{2+}-free medium causes the dissociation of tight junctions into discrete strands (Meldolesi et al., 1978). In inside-out follicles from pig thyroid gland, the tight junctions form an effective barrier against the exchange of surface markers. As discussed above, the tight junctions are by-passed by a vesicular shuttle between the apical and the basolateral cell surface domains. The restriction in membrane flow by tight junctions and the selectivity in the vesicular transfer of plasma membrane constituents are mechanisms which appear to operate in the maintenance of follicle cell polarity.

B. Release of Thyroglobulin from the Follicle Lumen in Pathological Conditions

It is well established that in most thyroid diseases the serum thyroglobulin levels are raised above normal concentrations (Table II) (Torrigiani *et al.*, 1969; van Herle *et al.*, 1979). In pathological conditions, mechanisms of thyroglobulin release distinct from transcytosis may predominate.

When rats are kept on an iodine-deficient diet, increased levels of serum thyroglobulin are observed (van Herle *et al.*, 1975). This observation corresponds to the elevated thyroglobulin concentrations found in patients with goiter (Pezzino *et al.*, 1978). Freeze-fracture studies have revealed a reduced number of strands in the tight junctions of goitrous follicles which are assumed to become leaky to thyroglobulin (van Uijen and van Dijk, 1983). In addition, leakage of thyroglobulin through defects in the follicle wall is observed when goiters become necrotic.

Diffusion of thyroglobulin from follicle lumina is also observed in patients with subacute thyroiditis which is characterized by inflammation with destruction of the follicle wall (Torrigiani *et al.*, 1969). Elevated concentrations are found in patients with differentiated thyroid carcinoma. The mechanism of thyroglobulin release is unknown but it has been speculated that invasion of normal tissue by the tumor causes leakage of thyroglobulin (van Herle *et al.*, 1979).

The vesicular transport across the follicle walls (transcytosis) and leakage of thyroglobulin caused by inflammation, malignant transformation, or relaxation of tight junctions in goitrous follicles are not contradictory. Both mechanisms, transcytosis and leakage, may operate in parallel and add up to the high concentrations of serum thyroglobulin observed in pathological conditions.

C. The Fate of Thyroglobulin Released into the Circulation

Thyroglobulin is a glycoprotein which contains about 10% carbohydrates (Spiro, 1965). It is well documented that glycoproteins in general are rapidly removed from the circulating blood plasma. Their elimination involves carbohydrate recognition and is primarily performed in the liver. The rapid clearance is explained by the preferential binding of glycoproteins to specific receptors, e.g., recognition of exposed galactosyl residues on desialated glycoproteins by hepatocyte membranes (Ashwell and Morell, 1974).

In the thyroglobulin molecule, the larger of two chains of oligosaccharides (subunit B) includes N-acetylneuraminic acid in terminal position (Yamamoto *et al.*, 1981). Desialation of thyroglobulin results in the exposure of galactosyl residues. In the rat, asialothyroglobulin is more rapidly removed from the circulating blood plasma (half-life about 15 minutes) than native thyroglobulin (half-life about 4.4 hours) (Ikekubo *et al.*, 1980). Hepatocytes play a key role in the clearance of glycoproteins and, most probably, also of thyroglobulin. Binding of serum thyroglobulin and endocytosis could give rise to the release of T3 and T4 by extrathyroidal degradation of thyroglobulin. Indeed, it has been observed that injection of radiolabeled thyroglobulin causes the appearance of thyroxine in the plasma of several mammalian species (Brown and Jackson, 1956).

VIII. Conclusions (Fig. 26)

Endocytosis from the apical plasma membrane of thyroid follicle cells is required for the lysosomal degradation of thyroglobulin and the release of thyroid hormones into the circulation. Inside-out follicles prepared from pig thyroid gland are a useful *in vitro* system because the apical plasma membranes are directed toward the culture medium and accessible for tracer studies on endo-

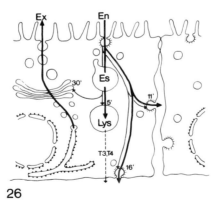

26

FIG. 26. Pathways of endocytosis (EN) in thyroid follicle cells as revealed with cationized ferritin (CF) (thin arrows) as a nonspecific membrane marker and with thyroglobulin (TG) (thick arrows) as a natural tracer. CF and TG share the endocytic pathways to endosomes (ES) and lysosomes (LYS, reached at ~5 minutes) and to the basal (reached at ~16 minutes) and the lateral (reached at ~11 minutes) cell surfaces. The transcytotic vesicles bypass the tight junctions and, probably, the lysosomes (LYS). In addition, CF (but not TG) is also transferred to stacked Golgi-cisternae (reached at ~30 minutes). It is unknown where both pathways separate. It is also unknown by which mechanisms T3 and T4 are released. In the Golgi complex, the endocytic pathway meets the export route (EX) of thyroglobulin.

cytosis. The endocytic pathways of thyroglobulin as a physiological ligand and of cationized ferritin (CF) as a nonspecific membrane marker are compared.

CF particles bind to the apical plasma membrane and collect in coated pits. Following stimulation with thyrotropin (TSH) the coated pits detach and form mainly smooth surfaced endocytic vesicles. CF particles appear first in endosomes and accumulate, begining at 5 minutes, also in lysosomes. Endosomes are characterized by their electron-lucent matrix, by the adherence of the tracer to the inner surface of the endosomal membrane, and by the absence of acid phosphatase activity. Endosomes are thought to be transformed into lysosomes which are distinguished from endosomes by their electron-opaque matrix, the detachment of CF from the lysosomal membrane, and the presence of acid phosphatase activity.

The membranes of endocytic vesicles, endosomes, and lysosomes do not show thyroperoxidase activity. This indicates that the thyroperoxidase-dependent iodination of thyroglobulin does not operate after endocytosis. It also suggests that endocytosis in thyroid follicle cells operates selectively with thyroperoxidase remaining in the apical plasma membrane.

Later, after $\sim$30 minutes of endocytosis, CF is also observed in stacked Golgi-cisternae indicating that part of the retrieved membrane is incorporated into Golgi membranes and, possibly, reutilized during the secretory cycle. In addition, a small fraction of endocytic vesicles becomes inserted by the formation of coated pits into the lateral (at $\sim$11 minutes) and the basal (at $\sim$16 minutes) plasma membranes. The vesicular transport across follicle cells (transcytosis) bypasses the tight junctions and, most probably, the lysosomal compartment. Transcytosis in thyroid follicle cells is temperature-sensitive and stopped at 15°C and below. The transcytotic activity is increased, however, about 9-fold when follicle cells are stimulated with TSH.

Thyroglobulin can be used as a tracer when conjugated to gold (ϕ 200 Å) or when radiolabeled with [^{3}H]leucine or with ^{125}I. The observations show that thyroglobulin is attached to the apical plasma membrane and transported mainly to endosomes and to lysosomes. A smaller fraction is also transferred across the follicle wall and released after the fusion of endocytic vesicles with the basolateral cell surface.

The transcytotic pathway and the route to endosomes and lysosomes are shared by both CF and thyroglobulin. Their endocytic transport differs, however, in that CF is also transferred to the stacked Golgi-cisternae which never contain thyroglobulin retrieved from the apical cell surface. It is concluded that the pathways for the endocytic membrane (as traced with CF) and for thyroglobulin separate from each other. It is still unknown where this separation takes place.

The studies with inside-out follicles confirm previous observations which showed that the main pathway of endocytosis in the thyroid is directed toward lysosomes. They also demonstrate that part of the endocytic membrane is recovered by being transferred to Golgi membranes and—on the transcytotic pathway—to the basolateral plasma membranes. Transcytosis in thyroid follicle cells explains the TSH-dependent appearance of thyroglobulin in the circulation of man and several mammalian species.

ACKNOWLEDGMENTS

I would like to thank Dr. F. Miller for reading this manuscript, Sabine Fuchs, Karin Hrubesch, Ulrike Reinhardt, and Doris Wolfram for technical assistance, Regina Schmittdiel for typing the manuscript, and Eva-Maria Praetorius for photographical work. The original research reported here was supported by Deutsche Forschungsgemeinschaft.

REFERENCES

Abrahamson, D. R., and Rodewald, R. (1981). *J. Cell Biol.* **91,** 270–280.
Ashwell, G., and Morell, A. G. (1974). *Biochem. Soc. Symp.* **40,** 117–124.
Bauer, W. C., and Meyer, J. S. (1964). *Science* **145,** 1431–1432.
Bernstein, L. H., and Wollman, S. H. (1976). *J. Ultrastruct. Res.* **56,** 326–330.
Björkman, U., Ekholm, R., and Denef, J.-F. (1981). *J. Ultrastruct. Res.* **74,** 105–115.
Brown, F., and Jackson, H. (1956). *Biochem. J.* **62,** 295–301.
Brown, M. S., Anderson, R. G. W., and Goldstein, J. L. (1983). *Cell* **32,** 663–667.
Burgess, D. R. (1982). *J. Cell Biol.* **95,** 853–863.
Chambard, M., Verrier, B., Gabrion, J., and Mauchamp, J. (1983). *J. Cell Biol.* **96,** 1172–1177.
Consiglio, E., Salvatore, G., Rall, J. E., and Kohn, L. D. (1979). *J. Biol. Chem.* **254,** 5065–5076.
Daniel, P. M., Pratt, O. E., Roitt, I. M., and Torrigiani, G. (1967). *Immunology* **12,** 489–504.
Danon, D., Goldstein, L., Marikowsky, Y., and Skutelsky, E. (1972). *J. Ultrastruct. Res.* **38,** 500–510.
Denef, J.-F., and Ekholm, R. (1980). *J. Ultrastruct. Res.* **71,** 203–221.
Denef, J.-F., Björkman, U., and Ekholm R. (1980). *J. Ultrastruct. Res.* **71,** 185–202.
Dragsten, P. R., Blumenthal, R., and Handler, J. S. (1981). *Nature (London)* **294,** 718–722.
Ekholm, R. (1981). *Mol. Cell. Endocrinol.* **24,** 141–163.
Ekholm, R., and Wollman, S. H. (1975). *Endocrinology* **97,** 1432–1444.
Ericson, L. E. (1981). *Mol. Cell. Endocrinol.* **22,** 1–24.
Ericson, L. E., and Engström, G. (1978). *Endocrinology* **103,** 883–892.
Farquhar, M. G. (1978). *J. Cell Biol.* **78,** R35–R42.
Farquhar, M. G. (1981). *Methods Cell Biol.* **23,** 399–472.
Farquhar, M. G., and Palade, G. E. (1963). *J. Cell Biol.* **17,** 375–412.
Farquhar, M. G., and Palade, G. E. (1981). *J. Cell Biol.* **91,** 77s–103s.
Friend, D. S., and Farquhar, M. G. (1967). *J. Cell Biol.* **35,** 357–376.
Fujita, H. (1975). *Inter. Rev. Cytol.* **40,** 197–280.
Gabrion, J., Travers, F., Benyamin, Y., Sentein, P., and van Thoai, N. (1980). *Cell Biol. Int. Rep.* **4,** 59–68.
Geuze, H. J., Slot, J. W., Strous, J. A., Lodish, H. F., and Schwartz, A. L. (1983). *Cell* **32,** 277–287.
Giraud, A., Gabrion, J., and Bouchilloux, S. (1981). *Exp. Cell Res.* **133,** 93–101.

Haddad, A., Smith, M. D., Herscovics, A., Nadler, N. J., and Leblond, C. P. (1971). *J. Cell Biol.* **49**, 856–882.

Haigler, H. T., McKanna, J. A., and Cohen, S. (1979). *J. Cell Biol.* **81**, 382–395.

Harding, C., Heuser, J., and Stahl, Ph. (1983). *J. Cell Biol.* **97**, 329–339.

Hayden, L. J., Shagrin, J. M., and Young, J. A. (1970). *Pflügers Arch.* **321**, 173–186.

Helenius, A., Marsh, M., and White, J. (1980). *Trends Biochem. Sci.* **5**, 104–106.

Helenius, A., Mellman, J., Wall, D., and Hubbard, A. (1983). *Trends Biochem. Sci.* **8**, 245–250.

Herzog, V. (1980). *In* "Biological Chemistry of Organelle Formation" (Th. Bücher, W. Sebald, and H. Weiss, eds.), pp. 119–145. 31st Colloquium der Gesellschaft für Biochemie, Mosbach.

Herzog, V. (1981). *Trends Biochem. Sci.* **6**, 319–322.

Herzog, V. (1983a). *J. Cell Biol.* **97**, 607–617.

Herzog, V. (1983b). *In* "Biomembranes. Part L: Membrane Biogenesis: Processing and Recycling" (S. Fleischer and B. Fleischer, eds.), Vol. 98, pp. 447–458. Academic Press, New York.

Herzog, V. (1983c). *Eur. J. Cell Biol. Suppl.* **I**, 21 (Abstr. No. 44).

Herzog, V., and Farquhar, M. G. (1977). *Proc. Natl. Acad. Sci. U.S.A.* **74**, 5073–5077.

Herzog, V., and Farquhar, M. G. (1983). *In* "Biomembranes. Part L: Membrane Biogenesis: Processing and Recycling" (S. Fleischer and B. Fleischer, eds.), Vol. 98, pp. 203–225. Academic Press, New York.

Herzog, V., and Miller, F. (1979a). *Eur. J. Cell Biol.* **19**, 203–215.

Herzog, V., and Miller, F. (1979b). *In* "Secretory Mechanisms," pp. 101–116. Society for Experimental Biology Symposium, Cambridge Univ. Press, London and New York.

Herzog, V., and Miller, F. (1980). *Eur. J. Cell Biol.* **22**, 195 (Abstract M 581).

Herzog, V., and Miller, F. (1981). *Eur. J. Cell Biol.* **24**, 74–84.

Herzog, V., and Reggio, H. (1980). *Eur. J. Cell Biol.* **21**, 141–150.

Hirokawa, N., Keller, Th. C. S., Chasan, R., and Mooseker, M. S. (1983). *J. Cell Biol.* **96**, 1325–1336.

Holtzman, E. (1976). *Cell Biol. Monogr.* **3**, 1–27.

Hopkins, C. R., and Trowbridge, I. S. (1983). *J. Cell Biol.* **97**, 508–521.

Hovsépian, S., Feracci, H., Maroux, S., and Fayet, G. (1982). *Cell Tissue Res.* **224**, 601–611.

Huxham, M., and Beck, F. (1981). *Cell Biol. Int. Rep.* **5**, 1073–1086.

Ikekubo, K., Pervos, R., and Schneider, A. B. (1980). *Metabolism* **29**, 673–681.

Izumi, M., and Larsen, P. R. (1978). *Metabolism* **27**, 449–460.

Korn, E. D. (1975). *MTP Int. Rev. Sci.* **2**, 1–26.

Leblond, C. P., and Bennett, G. (1977). *In* "International Cell Biology" (B. R. Brinkley and K. R. Porter, eds.), pp. 326–336. Rockefeller Univ. Press, New York.

Lissitzky, S. (1976). *Pharmacol. Ther. Part B* **2**, 219–246.

Meldolesi, J., Castiglioni, G., Parma, R., Nassivera, N., and De Camilli, P. (1978). *J. Cell Biol.* **79**, 156–172.

Miller, F., and Palade, G. E. (1964). *J. Cell Biol.* **23**, 519–552.

Nadler, N. J., Sarkar, S. K., and Leblond, C. P. (1962). *Endocrinology* **71**, 120–129.

Neary, J. T., Nakamura, Ch., Davidson, B., Soodak, M., Vickery, A. L., and Maloof, F. (1978). *J. Clin. Endocrinol. Metab.* **46**, 791–799.

Nitsch, L., and Wollman, S. H. (1980). *J. Cell Biol.* **86**, 875–880.

Novikoff, A. B., Novikoff, Ph. M., Ma, M., Shin, W.-Y., and Quintana, N. (1974). *Adv. Cytopharmacol.* **II**, 349–368.

Ottosen, P., Courtoy, P., and Farquhar, M. G. (1980). *J. Exp. Med.* **152**, 1–19.

Pastan, I., and Willingham, M. C. (1981). *Science* **214**, 504–509.

Pastan, I., and Willingham, M. C. (1983). *Trends Biochem. Sci.* **8**, 250–254.

Pezzino, V., Cozzani, P., Filetti, S., Galbiati, A., List, E., Squatrito, S., and Vigneri, R. (1977). *Eur. J. Clin. Invest.* **7**, 503–508.

Pezzino, V., Vigneri, R., and Squatrito, S. (1978). *J. Clin. Endocrinol. Metab.* **46,** 653–657.
Pisam, M., and Ripoche, P. (1976). *J. Cell Biol.* **71,** 907–920.
Quinton, P. M., and Philpot, C. W. (1973). *J. Cell Biol.* **56,** 787–796.
Rall, J. E. (1974). *Perspect. Biol. Med.* **17,** 218–226.
Renston, R. H., Jones, A. L., Christiansen, W. D., and Hradek, G. T. (1980). *Science* **208,** 1276–1278.
Rodewald, R. (1980). *J. Cell Biol.* **85,** 18–32.
Rodewald, R., Newman, S. B., and Karnovsky, M. J. (1976). *J. Cell Biol.* **70,** 541–554.
Roitt, I. M., and Torrigiani, G. (1967). *Endocrinology* **81,** 421–429.
Seljelid, R., Reith, A., and Nakken, K. F. (1970). *Lab. Invest.* **23,** 595–605.
Sly, W. S., and Stahl, P. (1978). *In* "Transport of Macromolecules in Cellular Systems" (S. C. Silverstein, ed.), pp. 229–244. Dahlem Konferenzen, Berlin.
Smeds, S. (1972). *Endocrinology* **91,** 1300–1306.
Spiro, M. J. (1965). *J. Biol. Chem.* **240,** 774–781.
Stein, O., and Gross, J. (1964). *Endocrinology* **75,** 787–798.
Steinman, R. M., Mellman, I. S., Muller, W. A., and Cohn, Z. A. (1983). *J. Cell Biol.* **96,** 1–27.
Straus, W. (1964). *J. Cell Biol.* **21,** 295–308.
Straus, W. (1983). *J. Histochem. Cytochem.* **31,** 78–84.
Strum, J. M., and Karnovsky, M. J. (1970). *J. Cell Biol.* **44,** 655–666.
Tartakoff, A. (1980). *Int. Rev. Exp. Pathol.* **22,** 227–251.
Thiele, J., and Reale, E. (1976). *Cell Tissue Res.* **168,** 133–140.
Tice, L. W., and Wollman, S. H. (1974). *Endocrinology* **94,** 1555–1567.
Tice, L. W., Wollman, S. H., and Carter, R. C. (1975). *J. Cell Biol.* **66,** 657–663.
Tice, L. W., Carter, R. C., and Wollman, S. H. (1976). *Endocrinology* **98,** 800–801.
Torrigiani, G., Doniach, D., and Roitt, I. M. (1969). *J. Clin. Endocrinol. Metab.* **29,** 305–314.
Tycko, B., and Maxfield, F. R. (1982). *Cell* **28,** 643–651.
Uller, R. P., van Herle, A. J., and Chopra, I. J. (1973). *J. Clin. Endocrinol. Metab.* **37,** 741–745.
van den Hove, M.-F., Couvreur, M., de Visscher, M., and Salvatore, G. (1982). *Eur. J. Biochem.* **122,** 415–422.
van Deurs, B., von Bülow, F., and Møller, M. (1981). *J. Cell Biol.* **89,** 131–139.
van Herle, A. J., Uller, R. P., Matthews, N. L., and Brown, J. (1973). *J. Clin. Invest.* **52,** 1320–1327.
van Herle, A. J., Klandorf, H., and Uller, R. P. (1975). *J. Clin. Invest.* **56,** 1073–1081.
van Herle, A. J., Vassart, G., and Dumont, J. E. (1979). *New Engl. J. Med.* **301,** 307–314.
van Uijen, A. J., and van Dijk, J. E. (1983). *Ann. Endocrinol.* **44,** 52A (Abstr. No. 88).
Wetzel, B. K., Spicer, S. S., and Wollman, S. H. (1965). *J. Cell Biol.* **25,** 593–618.
Whur, P., Herscovics, A., and Leblond, C. P. (1969). *J. Cell Biol.* **43,** 289–311.
Williams, J. A. (1969). *Am. J. Physiol.* **217,** 1094–1100.
Wolff, J. (1964). *Physiol. Rev.* **44,** 45–90.
Wollman, S. H. (1969). *In* "Lysosomes in Biology and Pathology" (T. Dingle and H. B. Fell, eds.), Vol. 2, pp. 483–512. North-Holland Publ., Amsterdam.
Wollman, S. H., and Loewenstein, J. E. (1973). *Endocrinology* **93,** 248–252.
Wollman, S. H., and Spicer, S. S. (1963). *In* "Thyrotropin" (S. C. Werner, ed.), p. 168. Thomas, Springfield, Illinois.
Wollman, S. H., Spicer, S. S., and Burnstone, M. S. (1964). *J. Cell Biol.* **21,** 191–201.
Yamamoto, K., Tsuji, T., Irimura, T., and Osawa, T. (1981). *Biochem. J.* **195,** 701–713.
Zeligs, J. D., and Wollman, S. H. (1977a). *J. Ultrastruct. Res.* **59,** 57–69.
Zeligs, J. D., and Wollman, S. H. (1977b). *J. Ultrastruct. Res.* **60,** 99–105.
Ziomek, C. A., Schulman, S., and Edidin, M. (1980). *J. Cell Biol.* **86,** 849–857.

INTERNATIONAL REVIEW OF CYTOLOGY, VOL. 91

Transport of Proteins into Mitochondria

SHAWN DOONAN,* ERSILIA MARRA,† SALVATORE PASSARELLA,†
CECILIA SACCONE,† AND ERNESTO QUAGLIARIELLO†

*Department of Biochemistry, University College, Cork, Ireland and †Istituto di Chimica Biologica e Centro di Studi sui Mitocondri e Metabolismo Energetico del Consiglio Nazionale delle Recerche, Università di Bari, Bari, Italy

I. Introduction

A. THE MITOCHONDRION

One of the main distinguishing features between prokaryotes and eukaryotes is the possession by the latter of specialized organelles responsible for the major oxidative functions of the organism. According to Klingenberg (1970), mitochondria constitute a "closed space" in which are contained the enzymes of terminal metabolism (citric acid cycle, terminal electron transfer chain, oxidative phosphorylation, β-oxidation, and, in part, urea synthesis and gluconeogenesis). The enzymatic capabilities of mitochondria vary somewhat from organism to organism, and indeed from tissue to tissue, as do the number and morphology of the organelles, but in spite of these differences the main features of mitochondrial function show remarkable consistency in organisms as diverse as yeast and mammals. As we will show in this review, the same consistency is also apparent, but with some differences, in the process by which mitochondria are biosynthesized.

For what follows, it will be necessary to distinguish between four structurally and functionally distinct regions of the mitochondrion, that is the two membranes which surround it, the space between these termed the intermembranal space, and the inner compartment or matrix. Some of the features of these regions are summarized below, largely following the review given by Chua and Schmidt (1979) which should be consulted for detailed references.

B. MITOCHONDRIAL COMPARTMENTS

1. External Membrane

This membrane contains relatively little protein constituting at most 4% of the total mitochondrial protein. The components so far identified include cytochrome b_5, monoamine oxidase, and kynurenine hydrolase. More recently a protein termed porin has been identified in the outer membrane of mitochondria from *Neurospora crassa* (Freitag *et al.*, 1982; Mannella, 1982), yeast (Mihara *et al.*, 1982b), and rat liver (Lindén *et al.*, 1982); this protein is thought to form ion-conducting channels through the membrane.

2. Intermembrane Space

Major enzymes present in this compartment include adenylate kinase, sulfite oxidase, cytocrome b_2 and cytochrome c peroxidase. Cytochrome c_1, which should properly be considered as a component of the inner membrane since a part of its polypeptide chain is inserted in that membrane, shares some biosynthetic features with cytochrome c peroxidase and cytochrome b_2 (see Section VII,A); consequently it will be referred to in what follows as an intermembranal protein.

3. *Inner Membrane*

This is a protein-rich membrane system containing about 20% of the total mitochondrial protein. This high protein content is accommodated by the extensive invaginations characteristic of the inner membrane. A major part (30–40%) of this protein is involved in terminal electron transport and ATP synthesis. The enzymes of the former process may be isolated as four complexes, namely NADH/coenzyme Q oxidoreductase (complex I), succinate/coenzyme Q oxidoreductase (complex II), coenzyme Q/cytochrome c oxidoreductase (complex III), and cytochrome oxidase (complex IV). An essential component of this system is cytochrome c which is a peripheral protein on the outer face of the inner membrane. Similarly, the proton-transducing ATPase (adenosine triphosphatase) consists of two parts, an integral complex F_0 and a peripheral complex F_1 which faces the matrix space of the mitochondrion.

The most abundant protein of the inner membrane is the carboxyatractyloside-binding protein whose function is the exchange of ADP and ATP across the inner membrane system; this protein has been isolated by Eiermann *et al.* (1977). In addition, a variety of other anion translocases exist but these will not concern us here.

4. *Matrix Space*

The matrix contains the remaining 60–70% of the mitochondrial protein and is the site of the major pathways such as the citric acid cycle and the β-oxidation. Other important enzymic activities from the present point of view are carbamoyl-phosphate synthetase, ornithine transcarbamylase, aspartate aminotransferase, glutamate dehydrogenase, malate dehydrogenase, ornithine aminotransferase, and manganese-dependent superoxide dismutase.

C. Scope of the Review

It is now established that mitochondria are produced by the concerted action of two genetic systems. The mitochondria accumulate proteins from both of these systems and ultimately divide, although the details of the latter process are far from clear. The major part of this review will be devoted to a summary of what is known about how proteins synthesized outside the mitochondria are translocated into one or other of the four compartments described above. Some aspects of this problem, such as the identities of proteins coded by the mitochondrial genome and of those coded by the nuclear genome, have been extensively reviewed previously (see, for example, Tzagoloff, 1982) and will be dealt with only briefly here. We will also refer to some of the characteristics of the processes by which proteins are exported from cells since, although there are some formal similarities between protein export and protein uptake into mitochondria, we believe that an overemphasis on these similarities may be misleading particularly with respect to the role played by N-terminal extensions of the precursors of mitochondrial proteins in uptake into the organelles.

Emphasis will also be placed on results obtained using a model system in which uptake of native proteins into mitochondria may be observed since this system has the advantage of ease of manipulation of experimental condition and allows the possibility of examination of the effects of chemical modifications on the uptake process. Finally we will attempt a synthesis of the current state of knowledge of protein transport into mitochondria. Previous reviews of these topics include those of Chua and Schmidt (1979), Schatz (1979), Waksman *et al.* (1980), Kreil (1981), Neupert and Schatz (1981), Sabatini *et al.* (1982), Ades (1982), Schatz and Butow (1983), and Hay *et al.* (1984).

II. Site of Coding of Mitochondrial Proteins

Early work in this area made use of the fact that mitochondrial protein synthesis is specifically inhibited by chloramphenicol whereas cytosolic protein synthesis is specifically inhibited by cycloheximide. Hence studies of protein synthesis in the presence of one or other of these compounds was of central importance in identifying which proteins are synthesized in the two compartments. In the case of yeast the availability of petite mutants lacking some or all of the mitochondrial DNA provided another experimental tool for identification of mitochondrially coded proteins. Using these methods, proteins coded by the mitochondrial genome in yeast were shown to include the three largest subunits of cytochrome oxidase, subunits VI and IX of the F_0-ATPase, and cytochrome *b*. A similar set of proteins is encoded by the mitochondrial genome of rat and of *Neurospora crassa* except that in these cases the subunit IX of F_0-ATPase is coded by the nucleus (Sebald, 1977; De Jong *et al.*, 1980). This early work has been extensively reviewed (see, for example, Tzagoloff *et al.*, 1979; Tzagoloff, 1982; Wallace, 1982) and will not be enlarged on here.

Dramatic confirmation of the results outlined above has been obtained as a result of sequence analysis on mitochondrial DNA from human (Anderson *et al.*, 1981), ox (Anderson *et al.*, 1982), rat (Pepe *et al.*, 1983), mouse (Bibb *et al.*, 1981), and yeast (see Rosamond, 1982, for references). In animal cells, the mitDNA is a closed circle of about 16,000 base pairs. It codes for the two ribosomal RNA molecules, a set of transfer RNAs, and also for the proteins listed above. In addition, there are eight open reading frames for which no protein product is known to date but, given the fact that the base sequences are highly conserved in these regions in the mitDNA from rat, ox, mouse, and human, they probably are involved in synthesis of proteins; these open reading frames have been termed URFs (for unassigned reading frames).

Although the mitochondrial genome of yeast is approximately five times larger than that of mammals its coding capacity seems not to be significantly greater. In addition to the proteins already mentioned, the mitDNA in yeast codes for one subunit (var 1) of the mitochondrial ribosome. A notable difference between the mitochondrial genes of yeast and mammals is that some of the former contain introns (genes for subunit I of cytochrome oxidase and for cytochrome b); there is evidence that one of these introns codes for a maturase involved in mitRNA splicing (Dujardin et al., 1982).

The picture that has emerged, then, is that the mitochondrial genome codes for five to six components of the inner mitochondrial membrane (plus var 1 in the case of yeast) and for a small number of components which are as yet unidentified and which, if they are indeed produced, are made in small quantity. Hence the vast majority of mitochondrial proteins, both in terms of number of species and of percentage of protein content of the organelles, are coded in the nucleus, synthesized in the cell cytosol, and subsequently imported into the organelles; it is generally accepted that import of mRNA from the cytosol and its translation within the mitochondria does not occur. This protein traffic is the subject of what follows.

III. Protein Export from Cells

The passage of proteins through membranes or insertion into membranes is obviously not unique to mitochondria but occurs also in such diverse situations as formation of secreted proteins by the liver, production of egg proteins in the avian oviduct, secretion of proteins by microorganisms and synthesis of certain viruses, as well obviously as in formation of all membranous structures within cells. Studies on some of these systems, particularly that involving transfer of proteins through the endoplasmic reticulum of mammalian cells, predated the study of protein import into mitochondria. Although this process is not of direct concern for the purpose of this review we feel it to be important to state briefly the main features of the process so that the similarities and differences between it and protein movement into mitochondria can be drawn subsequently. It is also the case, in our view, that there has been excessive emphasis on certain similarities between protein export from cells and import into mitochondria which may have impeded progress toward an understanding of the mechanism of the latter.

Protein export from cells and insertion into membranes have been extensively reviewed (Waksman et al., 1980; Wickner, 1980; Engelman and Steitz, 1981;

Kreil, 1981; Sabatini *et al.*, 1982; Strauss and Boime, 1982; Suominen and Mäntsälä, 1983) and only an outline is given here. The main features are as follows:

1. Proteins to be exported generally contain a predominantly hydrophobic presequence of 20 to 30 amino acid residues.

2. Synthesis starts on free cytosolic ribosomes. At a stage when the presequence has emerged from the large ribosomal subunit, attachment to the endoplasmic reticular membrane occurs.

3. The presequence passes through the endoplasmic reticular membrane as further protein synthesis occurs and the presequence is cleaved by a protease (signalase) on the lumenal face of the membrane.

4. Completion of protein synthesis and passage through the membrane systems, followed or accompanied by protein folding, produce the mature protein in the lumen of the endoplasmic reticulum.

Knowledge of step 2 has recently been refined by the work of Blobel and his colleagues (Gilmore *et al.*, 1982a,b) who have demonstrated the existence of a cytosolic signal recognition particle (SRP), composed both of protein and a 7 S RNA, whose function is to bind to the presequence (signal) emerging from the large ribosomal subunit and to promote interaction with the endoplasmic reticular membrane, this interaction being mediated by a docking protein in the membrane.

It must be emphasized that although certain aspects of protein export are now reasonably well established, the critical events involving passage of protein through the reticular membrane are still obscure and a matter of debate. There is no concensus of opinions as to whether transport occurs through the lipid phase of through a proteinaceous pore (although the recent work by Blobel and co-workers outlined above may argue for the latter possibility). It is also unknown whether the presequence transits the membrane as single polypeptide chain or as a double strand as proposed in the helical hairpin model of Engelman and Steitz (1981). Recent work on ovalbumin (Meek *et al.*, 1982), a major secretory protein which lacks a terminal presequence, has lent support to the hairpin model; in this case a sequence of amino acids between positions 25 and 45 constitutes the membrane recognition and insertion signal, and furthermore this sequence could form a helical hairpin with the N-terminal section of the protein.

Finally, the source of energy required to move proteins across membranes is not clear. This could be provided by a "push" resulting from continuing peptide synthesis on a ribosome which is firmly anchored to the membrane, or from a "pull" resulting from folding of that part of the protein already translocated, or from both of these.

Topogenesis of integral membrane proteins could occur by a variant of the mechanism described above. If, during the process of transit a very hydrophilic stretch of amino acids is produced, then this might prevent further trans-location leaving the C-terminal region of the protein on the cytosolic side of the membrane and resulting in a protein with a single stretch of amino acids crossing the membrane. Alternatively, the C-terminal section may contain one or more internal transit signals resulting in loops of polypeptide transversing the membrane. Again these possibilities have been reviewed (Engelman and Steitz, 1981; Sabatini *et al.*, 1982) and will not be treated in detail here.

IV. Biosynthesis of Mitochondrial Proteins in the Cytosol

In the light of the discussion in Section III some obvious questions arise concerning the mechanism of biosynthesis of nuclearly-coded mitochondrial proteins.

1. What is the site of synthesis of such proteins; that is, are they synthesized on free cytosolic ribosomes or on ribosomes associated with some membrane system?

2. What is the timing of translocation into or through mitochondrial membranes; is this a cotranslational or posttranslational event?

3. Are these proteins synthesized as the native forms or as precursors with molecular weights or conformations different from those of the native proteins?

These questions are discussed in the present section before turning to the problem of how translation products are translocated into mitochondria.

A. SITE OF SYNTHESIS

Early work on this problem, reviewed by Shore and Tata (1977a,b,c), and by Ades (1982) suggested that mitochondrial proteins are synthesized by polyribosomes bound to membranous structures. Suggestions that the rough endoplasmic reticulum is the site of synthesis came from work on cytochrome *c* (Gonzalez-Cadavid and de Cordova, 1974), malate dehydrogenase (Bingham and Campbell, 1972), and glutamate dehydrogenase (Godinot and Lardy, 1973). For example, in the last mentioned case, rats were injected with radiolabeled amino acids and their livers subsequently subjected to fractionation and localization of labeled glutamate dehydrogenase using immunochemical methods. Substantial amounts of labeled enzyme were found in the rough microsomal fraction. There is, however, a difficulty in interpretation of results of this kind since it is likely that enzymes and other proteins will redistribute during the process of subcellular fractionation. We have shown (Marra *et al.*, 1977a), for example, that purified mitochondrial aspartate aminotransferase has a high affinity for microsomes in solutions of low ionic strength such as the sucrose media frequently used for subcellular fractionation.

Another possibility was proposed by Butow and his collaborators working with yeast (Kellems and Butow, 1972; Kellems *et al.*, 1975). Electron micrographs of actively metabolizing yeast cells showed extensive binding of ribosomes to the cytosolic face of the outer membranes of the mitochondria, particularly in regions where the inner and outer membranes appeared to be in close juxtaposition. Moreover, treatment of such mitochondria with puromycin led to discharge of nascent polypeptides into the organelles. These observations suggested a close similarity between the process of transport of proteins through the endoplasmic reticulum and import into mitochondria, the latter being viewed as a cotranslational vectorial transport. More recent work, to be reviewed at a later stage (Section IV,B), suggests that such a mechanism may occur in yeast under certain conditions but that it is not an essential part of the protein uptake process. In other organisms, no evidence has been obtained for mitochondrially associated cytoplasmic ribosomes.

The weight of more recent experimental evidence shows that mitochondrial proteins are synthesized on free cytosolic ribosomes. The usual method for demonstration of this fact is to isolate the various classes of polysomes by subcellular fractionation and to establish which of them carry mRNA coding for particular mitochondrial proteins; this is done by translation in cell-free systems followed by specific immunoprecipitation. Synthesis on free polysomes has been established for proteins destined for each of the mitochondrial compartments. A partial list is as follows.

Outer membrane: porin from *Neurospora crassa* (Freitag *et al.*, 1982); 35K protein from rat liver (Shore *et al.*, 1981). Intermembrane space: sulfite oxidase from rat liver (Ono *et al.*, 1982; Mihara *et al.*, 1982a).

Inner membrane: β-hydroxybutyrate dehydrogenase from rat liver (Mihara *et al.*, 1982a); ADP/ATP carrier from *Neurospora crassa* (Zimmermann and Neupert, 1980); subunit IX of the F_0-ATPase from *Neurospora crassa* (Schmidt *et al.*, 1983).

Matrix: aspartate aminotransferase from chicken heart and rat liver (Sonderegger *et al.*, 1982; Sakakibara *et al.*, 1980); ornithine aminotransferase from rat liver (Hayashi *et al.*, 1981); glutamate dehydrogenase from rat liver (Mihara *et al.*, 1982a); malate dehydrogenase from rat liver (Mihara *et al.*, 1982a).

In summary then the picture is now emerging that, except in yeast under particular metabolic conditions, proteins destined for import into all four mitochondrial compartments are synthesized on free cytosolic ribosomes. This is clearly very different from the situation with protein export from cells.

B. Timing of Translocation

The demonstration that, for many if not all proteins and tissues the site of synthesis of mitochondrial proteins is predominantly free cytosolic robosomes is in itself an indication that transport into mitochondria is a posttranslational event. If binding of the nascent polypeptide chain to mitochondria occurred even as a late event in synthesis then a substantial proportion of polysomes carrying messages for mitochondrial proteins would be found associated with the organelles which is contrary to experience (with the exception of yeast; see below). Hence the more likely sequence of events is that completed translation products are released into the cytosol and subsequently imported into the mitochondria.

A direct experimental demonstration that this is so was given in an important publication from Neupert's laboratory (Hallermayer *et al.*, 1977). In this work, *Neurospora* cells were labeled with ^{35}S, then pulsed with [^{3}H]leucine and chased with unlabeled leucine. Subsequent subcellular fractionation showed that the radiolabel appeared in mitochondria more slowly than in other subcellular fractions. More importantly, when mitochondrial proteins were immunoprecipitated labeled products appeared first in the cytosol and only much more slowly in the mitochondria; addition of cycloheximide, which prevented appearance of new labeled proteins in the cytosol, did not prevent further accumulation of label into the mitochondria. Moreover, the kinetics of appearance of different proteins in the mitochondrial fraction were not the same. These results provided the first compelling evidence that uptake of proteins into mitochondria is not coupled to protein synthesis and moreover that completed translation products are released into cytoplasmic pools before import.

More recently Ades and Butow (1980a,b) have published similar work using yeast, the results of which seem to be partially in conflict with those of Neupert's group. In the first of these two papers Ades and Butow return to the question discussed in Section IV,A concerning the role of cytoplasmic ribosomes bound to the outer face of the mitochondria and show that, in a read-out system, 80% of the products of these polysomes are imported into mitochondria. Moreover, they show that the proportion of the α, β, and γ subunits of F_1-ATPase synthesized by membrane-bound ribosomes is greater than the proportion synthesized by free ribosomes. In the second publication (Ades and Butow, 1980b) a double-isotope pulse-labeling technique was used to determine the kinetics of labeling of mitochondrial proteins and of a cytosolic protein. No lag period was found for labeling of mitochondrial proteins compared with cytosolic proteins; this argued against the existence of detectable pools of mitochondrial proteins in the cell cytosol awaiting import. When synthesis was stopped with cycloheximide the decrease in incorporation of label into cytosolic and mitochondrial proteins occurred at equal rates. However, when cold methionine was added as a chase during labeling, incorporation of label into mitochondria persisted with a half-life of 1.5 minutes whereas incorporation into cytosolic proteins did not. More-over, in a synchronized restart of protein synthesis after ribosome run-off, the initial rate of incorporation of label into cytosolic proteins was greater than into mitochondrial proteins.

Taken together, these results suggest that in yeast under the conditions em-ployed by Ades and Butow, the pool of completed translation products is too small to be detected under steady-state conditions. Moreover, the binding sites for import of mitochondrial proteins (see Section VI,E) are able to recognize and bind partially completed translation products. But as the authors recognize, this does not mean that ribosome–membrane interaction nor polypeptide chain elongation are important in the transport process. Most likely, the phenomenon observed with yeast (and not with other organisms) reflects a different balance between the rate of protein synthesis and rates of import into the organelles; if the latter is more rapid than the former, then pool sizes will be small and binding sites available for interaction with partially synthesized proteins. It is likely however, that proteins from both sources, those completed and free in the cytosol and those undergoing completion by ribosomes associated with the mitochon-dria, will be imported by the same mechanism.

In a recent paper, Suissa and Schatz (1982) have also examined the contribu-tion of membrane-bound polysomes in yeast to the synthesis of proteins destined for all four mitochondrial compartments. They find that, although the polysomes bound to the outer face of the mitochondria carry messenger RNAs specifically for mitochondrial proteins, the percentage of the mRNA for these proteins in mitochondria-associated polysomes is generally smaller than the percentage of mRNA in free cytosolic polysomes and in no case is the former predominant. Further, Reid and Schatz (1982b) have demonstrated that extramitochondrial pools of precursor proteins are formed in yeast cells under appropriate conditions *in vivo,* and that these precursors can be imported posttranslationally into mitochondria. These results are consistent with the view that the major part of mitochondrial protein synthesis in yeast occurs by liberation of completed trans-lation products into the cytosol followed by import after translation and that a relatively minor role is played by membrane-associated ribosomes.

Perhaps the most definitive evidence for the completion of translation of proteins before import is provided by the wealth of experiments where mitochon-drial proteins synthesized in cell-free systems have been imported into mitochon-dria subsequently added to the system. This topic is, however, more convenient-ly discussed in Section V,A.

C. Precursors of Mitochondrial Proteins

Up to this point, comparison of protein uptake into mitochondria with protein export from cells has shown only differences between the two processes. There is, however, one similarity at least for the majority of mitochondrial proteins, which is that they are initially synthesized as precursors of higher molecular weight than that of the native forms. The examples of this that are now established are too numerous to describe in detail. Table I shows some representative examples covering a variety of organisms and of submitochondrial locations of the particular proteins. In nearly all cases, the technique used was translation of messenger RNA in a cell-free system followed by immunoprecipitation and analysis of the size of the product on SDS–polyacrylamide gel electrophoresis. More recently, precursors of higher molecular weight than the mature forms have been detected *in vivo* using cell cultures. Examples include aspartate aminotransferase in chicken embryo fibroblasts (Jaussi *et al.*, 1982), and ornithine transcarbamylase and carbamoyl-phosphate synthetase in rat hepatocytes (Mori *et al.*, 1981a); other examples will be described in Section V,B.

The assumption is generally made that the extra piece of these precursors is N-terminal and, although this is a reasonable assumption, it has been verified in relatively few cases. Initial rather indirect evidence on this point came from experiments in which cell-free synthesis of precursors was carried out in the presence of N-formyl-[^{35}S]methionine-tRNA$_i$ under conditions where no labeled methionine would be incorporated internally into the translation products. Label was incorporated into precursors (since mammalian systems lack a formylase) and the label was lost on subsequent uptake of precursors into mitochondria, suggestive of N-terminal processing. Examples of proteins studied by this method include ornithine transcarbamylase from rat liver (Kraus *et al.*, 1981), F$_1$-ATPase from yeast (Lewin *et al.*, 1980), and cytochrome oxidase from yeast (Lewin *et al.*, 1980; Mihara and Blobel, 1980). The last-mentioned example is interesting in that the individual subunits of cytochrome oxidase were labeled which would not have been the case if, as claimed by Poyton and McKemmie (1979a,b), the initial translation product is a polyprotein containing all of the cytoplasmically translated subunits of the enzyme.

Direct demonstration of an N-terminal extension has been provided by partial amino acid sequence analysis in the case of aspartate aminotransferase (Sannia *et al.*, 1983); the prepiece consists of a sequence of 24 amino acid residues. More complete information is available in two cases from nucleotide sequencing. Viebrock *et al.* (1982) have sequenced cDNA clones of the mRNA for the proteolipid subunit IX of the ATPase from *Neurospora crassa*. They find an N-terminal prepiece of 66 amino acid residues with a predominantly polar character; 12 of the amino acids are basic. Kaput *et al.* (1982) have isolated and sequenced the genomic DNA for yeast cytochrome c peroxidase. The presequence consists of 68 amino acid residues; again the presequence contains a considerable number of basic residues but also a stretch of 23 nonpolar residues.

More recently, Nagata *et al.* (1983) have isolated and sequenced a nuclear gene fragment from yeast coding for the mitochondrial elongation factor Tu. The

TABLE I

SOME REPRESENTATIVE EXAMPLES OF MITOCHONDRIAL PROTEINS SYNTHESIZED AS
PRECURSORS OF HIGHER MOLECULAR WEIGHT THAN THE NATIVE FORM

Mitochondrial compartment	Source	Protein	References
Intermembranal space	Rat liver	Sulfite oxidase	Ono and Ito (1982); Mihara et al. (1982a)
	Yeast	Cytochrome c peroxidase	Maccecchini et al. (1979b)
	Yeast	Cytochrome b_2	Gasser et al. (1982a)
Inner membrane	Rat liver	Cytochrome oxidase subunit IV	Schmelzer and Heinrich (1980)
	Yeast	Cytochrome oxidase subunits IV, V, and VI[a]	Lewin et al. (1980); Mihara and Blobel (1980)
	Neurospora crassa	ATPase subunit IX	Schmidt et al. (1983a); Michel et al. (1979)
	Sweet potato	Succinate dehydrogenase	Hattori et al. (1983)
	Rat liver	β-Hydroxybutyrate dehydrogenase	Mihara et al. (1982a)
	Dog heart	Creatine kinase	Perryman et al. (1983)
	Yeast	Cytochrome bc_1 subunit V	Côté et al. (1979)
	Neurospora crassa	Cytochrome bc_1 subunits I, II, IV, V, VII, and VIII	Teintze et al. (1982)
Matrix	Yeast	Citrate synthase	Alam et al. (1982)
	Neurospora crassa	Citrate synthase	Harmey and Neupert (1979)
	Rat liver	Glutamate dehydrogenase	Miralles et al. (1982); Mihara et al. (1982a)
	Rat liver	Malate dehydrogenase	Aziz et al. (1981); Mihara et al. (1982a)
	Yeast	Manganese superoxide dismutase	Autor (1982)
	Yeast	RNA polymerase	Lustig et al. (1982)
	Rat liver	Carbamoyl-phosphate synthetase	Shore et al. (1979); Mori et al. (1979a)
	Rat liver	Ornithine transcarbamylase	Mori et al. (1980); Conboy et al. (1979)
	Rat liver	Serine pyruvate aminotransferase	Oda et al. (1981)
	Rat liver	Aspartate aminotransferase	Sakakibara et al. (1980)
	Pig heart	Aspartate aminotransferase	Sannia et al. (1982)
	Chicken heart	Aspartate aminotransferase	Sonderegger et al. (1980)

TABLE I   (*Continued*)

Mitochondrial compartment	Source	Protein	References
	Chicken liver	δ-Aminolevulinate synthase	Ades and Harpe (1981)
	Rat liver	δ-Aminolevulinate synthase	Yamauchi *et al.* (1980)
	Yeast	F_1-ATPase α, β, and γ subunits	Macceccini *et al.* (1979a)

[a]Claims that the subunits of cytochrome oxidase from yeast are synthesized as a single polyprotein translation product (Poyton and McKemmie, 1979a,b) have subsequently been shown to be unfounded (see Section IV,C).

gene codes for a protein of 437 amino acid residues with obvious homology to the enzyme from *E. coli* but 37 residues longer at the N-terminus. This N-terminal section may represent a presequence of the mature protein and it is of interest that it shows structural similarity to the presequence of cytochrome *c* peroxidase (see above). It seems likely that, in the near future, sequencing of genomic DNA or of cDNA synthesized from messengers for mitochondrial proteins will provide a considerable amount of information on presequences and on other aspects of protein import into mitochondria. Proteins currently under study in this context include ornithine transcarbamylase from rat liver (Horwich *et al.*, 1983), subunit V of cytochrome oxidase from yeast (Cumsky *et al.*, 1983), subunits of complex 111 from yeast (van Loon *et al.*, 1982), tyrosine aminotransferase from rat liver (Scherer *et al.*, 1982), the β-subunit of ATPase from yeast (Saltzgaber-Muller *et al.*, 1983), and a protein involved in processing of transcripts of the yeast mitochondrial cytochrome *b* gene (McGraw and Tzagoloff, 1983).

Several proteins are also known for which the initial translation product has the same size as the mature protein. A recent example is the protein porin from the outer membrane (Freitag *et al.*, 1982; Mihara *et al.*, 1982b; Gasser and Schatz, 1983) but this protein may perhaps be considered untypical of mitochondrial proteins since its assembly into the outer membrane is not energy dependent (see Section VI,B). In the case of ornithine aminotransferase, an enzyme of the matrix, there is dispute on this point; Hayashi *et al.* (1981) do not find a precursor for the rat liver enzyme but Mueckler *et al.* (1982) find a precursor with a molecular weight approximately 6000 greater than that of the native form. More recently it has been shown by two groups (Suissa and Schatz, 1982; Hampsey *et al.*, 1983) that the 2-isopropylmalate synthase of yeast, a matrix protein, is synthesized in cell-free systems in a form with molecular weight indistinguishable from that of the native enzyme. Hampsey *et al.* (1983) showed that the protein was labeled when made in the presence of *N*-formyl-[^{35}S]methionine-tRNA$_i$ but did not determin whether the label was retained when the protein was taken up into mitochondria. The possibility remains, therefore, that 2-isopropylmalate synthase is synthesized with a very short prepiece such that there is no detectable difference in size between precursor and mature protein as judged by electrophoretic analysis. Hence it remains to be established unambiguously that any matrix protein is synthesized without at least a short extension. In the case of enzymes of the intermembranal space, it has been claimed that adenylate kinase is synthesized in its native form (Watanabe and Kubo, 1982).

A more detailed study has been carried out by Neupert's group on cytochrome *c* and on the ADP/ATP translocator protein from *Neurospora crassa* (Korb and Neupert, 1978; Zimmermann and Neupert, 1980). In neither case is the protein synthesized as a precursor of higher molecular weight than the native form; there is, however, evidence that the newly synthesized protein in both cases differs in conformation from the native form. Hence both of these newly synthesized proteins are described as precursors but the term is used in a different sense from the normal usage. In the case of cytochrome *c,* a difference in conformation is not surprising since the linkage of heme to the apoprotein to form holocytochrome *c* is likely to lead to such a change; the linkage of the heme occurs after transport to the outer face of the inner mitochondrial membrane (Korb and Neupert, 1978). Direct demonstration of this difference in conformation was provided by immunochemical results. Antibodies raised to chemically prepared apocytochrome *c* did not precipitate holocytochrome *c* but did precipitate the newly synthesized protein. Similarly antibodies to holocytochrome *c* did not precipitate apocytochrome *c.* Hence the evidence suggests a close similarity between newly synthesized and chemically prepared apocytochrome *c* and a conformational difference between these and holocytochrome *c.* For the ADP/ATP translocator protein (Zimmermann *et al.,* 1979; Zimmermann & Neupert, 1980) the evidence again suggests that the newly synthesized protein has the authentic molecular weight but differs in conformation from the mature form. The evidence for the latter point was that gel filtration and sedimentation analysis showed the protein to exist in a water-soluble multimeric form; since the mature translocator is an integral membrane protein and insoluble in aqueous solution then this implies a conformational difference compared with the newly synthesized form. Interestingly, in a recent report by Chien and Freeman (1983) it is claimed that the translocator protein is synthesized as a precursor with molecular weight greater than that of the mature protein in reticulocyte lysates primed with rat liver mRNA; this observation underlines the possibility that details of the biosynthesis of mitochondrial proteins may vary from species to species. The same group (Freeman *et al.,* 1983) have also shown that the inner membrane protein variously described as 32,000 M_r protein, uncoupling protein or thermogenin from rat brown adipose tissue is synthesized with a molecular weight indistinguishable from that of the native protein.

In summary, with few exceptions, mitochondrial proteins appear to be synthesized as precursors with higher molecular weights than the native forms or, when this is not so, with different conformations from those of the native forms. Clearly, in those cases when the precursor has an extra sequence of amino acids the conformation is also likely to be different from that of the native proteins. A recent example where this has been shown to be the case is with ornithine transcarbamylase (Mori *et al.*, 1982).

D. CYTOSOLIC ISOENZYMES AS "PRECURSORS" OF MITOCHONDRIAL FORMS

The supposition is normally made that precursor forms of mitochondrial proteins which accumulate in the cell cytosol are catalytically inactive and play no metabolic role until after transport into the mitochondria. There is some recent evidence that this is not always so.

The enzyme malonyl-CoA decarboxylase is normally present only in the mitochondria of animal cells but in the specialized uropygial gland of water fowl it is found in large amounts in the cytosol (Kim and Kolattukudy, 1978a,b). Recent evidence suggests (Flurkey *et al.*, 1982) that the cytosolic and mitochondrial forms of the enzyme are identical except that the latter has a molecular weight about 3000 smaller than the former. Hence the cytosolic isoenzyme may be considered as the "precursor" of the mitochondrial form but in this case the precursor, by virtue of its catalytic activity, must be considered to be identical or very similar in conformation to the mitochondrial enzyme.

A similar example, but one of more widespread occurrence not involving a highly specialized tissue, is provided by fumarase. Some time ago, Edwards and Hopkinson (1979) obtained genetic evidence that the cytosolic and mitochondrial fumarases are coded by the same gene. The structural relationship between these two proteins was therefore of obvious interest and has been the subject of recent studies (O'Hare and Doonan, 1984). The two proteins have very similar molecular weights such that no difference is observable on SDS–polyacrylamide gel electrophoresis. Peptide maps show that most of the polypeptide chain is common to both isoenzymes but the cytosolic isoenzyme yielded two small peptides not found for the mitochondrial form. N-terminal amino acid analysis yielded alanine for the mitochondrial isoenzyme and glutamic acid/glutamine for the cytosolic form. The results strongly suggest that the mitochondrial isoenzyme is produced from the cytosolic form by removal of a small N-terminal fragment. Confirmation of this awaits N-terminal sequence analysis of the two proteins.

Genetic evidence for this phenomenon has also been provided in the case of two yeast enzymes involved in tRNA methylation (Hopper *et al.*, 1982). Yeast mutants lacking one or other of these two enzymes failed to methylate both cytosolic tRNA and mitochondrial tRNA; that is, the phenotype of the mutants was the same at the level of whole-cell tRNA and mitochondrial tRNA. This again suggests that a single enzyme is responsible for each of these methylase activities in the cytosol and in the mitochondria. No structural information is, as yet, available for these enzymes.

A final example of this type of relationship between cytosolic and mitochondrial proteins may be provided by δ-aminolevulinate synthase from rat liver (Nakakuki *et al.,* 1980; Yamauchi *et al.,* 1980; Kikuchi and Hayashi, 1981). Treatment of animals with allylisopropylacetamide leads to accumulation of a large amount of this enzyme in the cell cytosol. The form found in the cytosol is larger than that found in the mitochondria (Nakakuki *et al.,* 1980) and appears to be the precursor in transit to its mitochondrial location. Again this "precursor" form is catalytically active, but in this case, in distinction from the other two discussed above, the precursor form may not have a physiological role in the cytosol.

E. IDENTICAL ISOENZYMES IN THE TWO COMPARTMENTS

No clear-cut evidence for the existence of identical proteins in the cytosol and the mitochondria of the same cell has yet been provided. A claim that this is the case for phosphoenolpyruvate carboxykinase in bullfrog liver (Goto *et al.,* 1981) was based on insubstantial evidence and has not confirmed. More recently, the avian isoenzymes have been shown to be translated from different mRNA species, the mitochondrial isoenzyme as a precursor of molecular weight 2000 greater than that of the mature form (Hod *et al.,* 1982). This casts some doubt on the work with the isoenzymes from bull frog, although the possibility of species differences cannot be ruled out (see Section IV,C).

V. Systems for Study of Protein Import

Some of the features of biosynthesis of mitochondrial proteins can now be considered fairly firmly established, that is, which proteins are made in the cell cytosol, the site of translation of the mRNAs for these proteins, and the fact that the majority of such proteins are made in some sort of precursor form. What are, from many points of view, the most interesting aspects of the process, including specific recognition of proteins by the mitochondria and the events occurring during translocation across the mitochondrial membrane or membranes, remain largely obscure. It is with these aspects that the remainder of this review is concerned.

Before proceeding, however, it is necessary to outline the experimental techniques that have been used to obtain information about these processes since in this, as in other fields of investigation, availability of appropriate experimental methods is the essential prerequisite for progress to be made.

A. IMPORT OF PRECURSORS SYNTHESIZED IN CELL-FREE SYSTEMS

This may now be considered the classical technique in studies of protein import into mitochondria and had its origins in the pioneering work of the groups of Schatz, Neupert, Mori, and others. Early examples of this type of work include studies on the F_1-ATPase (Maccecchini *et al.*, 1979a), the ADP/ATP translocator protein (Zimmermann and Neupert, 1980), and carbamoyl-phosphate synthetase (Mori *et al.*, 1979b). The method used by all these groups, and by many others subsequently, involves synthesis of the precursor protein in a cell-free system programmed with RNA from the organism or tissue of interest, followed by incubation of the cell-free system with mitochondria for an extended period of time. Subsequently, mitochondria are ruptured and the protein of interest immunoprecipitated and analyzed, usually by electrophoresis on SDS–polyacrylamide gel. In those cases (the majority) where uptake is accompanied by proteolytic processing of the precursor, the analytical tool used is the appearance in mitochondria of radiolabeled protein, specifically precipitated with antiserum, and of the same molecular weight as the mature protein but of a smaller molecular weight than the precursor form. This last-mentioned characteristic is not, of course, observable for those proteins made without an extra piece of amino acid sequence.

The number of studies reported using this technique is now very large and we will not attempt to give a complete bibliography. Examples showing points of specific interest will be described in Section VI. Recently it has been claimed that import and processing of the precursor or ornithine transcarbamylase in systems of this type require a component or components (possibly proteinaceous) derived from the reticulocyte lysate used for synthesis of the precursor (Argan *et al.*, 1983; Miura *et al.*, 1983). Although of interest the significance of these results is as yet by no means clear and they will not be referred to further here.

B. IMPORT OF PRECURSORS SYNTHESIZED *in Vivo*

As an extention of the method described above, several reports have now appeared using unicellular organisms or animal cells in culture, to demonstrate the synthesis of precursors and their subsequent sequestration and processing by mitochondria *in vivo*.

For example, Reid and Schatz (1982a,b) have studied the biosynthesis of cytochrome b_2 and cytochrome *c* peroxidase in intact yeast cells. Cells pulse-labeled in the presence of carbonylcyanide *m*-chlorophenylhydrazone (CCCP), which blocks import of proteins into mitochondria (see Section VI,B), accumulate precursors to both enzymes; this was shown by immunoprecipitation followed by electrophoresis on SDS–polyacrylamide gels. A subsequent chase after removal of CCCP with excess 2-mercaptoethanol resulted in import and processing of the precursors as shown by the appearance of labeled proteins of the mature size. After pulses of short duration precursors were detectable even in the absence of CCCP, and their subsequent fate could be followed during a chase (Reid and Schatz, 1982b).

Similar experiments have been done with animal cells in culture. For example Mori *et al.* (1981a) have shown that pulse-labeled hepatocytes accumulate low levels of precursors of ornithine transcarbamylase and carbamoyl-phosphate synthetase; these precursors may be chased into mitochondria as shown by the disappearance of precursors from the cytosol. Jaussi *et al.* (1982) used chick embryo fibroblasts in their work on mitochondrial aspartate aminotransferase. Again, CCCP prevented processing of the precursor and led to its accumulation in the cells; in a subsequent chase in the presence of cysteamine, processing to the mature form was observed. The analytical technique employed was again SDS–polyacrylamide gel electrophoresis.

From these examples it can be seen that the essential features of the methods used are the same as those for uptake experiments *in vitro* except that, in some cases, it is necessary to initially block processing by use of an inhibitor. The value of the use of whole cells is, of course, that experiments are carried out under conditions approaching those operational *in vivo*. Interestingly, as will be described later (Section VI,C) some critical differences are observed when uptake *in vitro* and *in vivo* are compared.

C. MODEL SYSTEMS USING MATURE ENZYMES

The experimental systems described above suffer from some disadvantages. For example, it has not yet proved possible to isolate precursors in sufficient quantity so that chemical modification studies can be carried out to probe the structural requirements for protein uptake; an exception is the case of cytochrome *c* where the chemically prepared apo-protein is functionally equivalent to the precursor. In addition, cell-free systems are not well defined nor easy to manipulate in some respects, so that it is not easy to examine the parameters which govern protein uptake into mitochondria using these systems; the same problem applies to experiments conducted with intact cells.

To overcome these difficulties we have worked extensively with a model system in which import of purified mature mitochondrial isoenzymes into mitochondria can be observed. It is necessary to describe these systems in some detail, and in particular to review the evidence that the phenomena observed are indeed due to protein import into mitochondria, since there have been some claims (for example Harmey *et al.*, 1977; Sakakibara *et al.*, 1981; Kraus *et al.*, 1981) that mature proteins cannot be so imported.

The initial work was done with rat liver mitochondrial aspartate aminotransferase and two methods were devised to measure uptake of the isoenzyme into isolated organelles. The first method (Marra *et al.*, 1977b) involved measurement of intramitochondrial enzyme activity and then measurement of the increase in this activity after incubation of the organelles with purified aspartate aminotransferase. The principles of the assay method are as follows. The mitochondria are treated with rotenone and sodium arsenite to prevent oxidation of intramitochondrial NADH by electron transport or by 2-oxoglutarate dehydrogenase. Aspartate (one of the substrates of the enzyme) is allowed to accumulate inside the mitochondria after which 2-oxoglutarate is added. Import of the latter via its carrier provides the substrate pair for the enzyme, and action of aspartate aminotransferase produces oxaloacetate as one product. This is reduced by intramitochondrial NADH catalyzed by endogenous malate dehydrogenase resulting in oxidation of the NADH, the observed parameter being the decrease in fluorescence of the latter substance. The rate of decrease of fluorescence provides a true measure of the rate of action of intramitochondrial aspartate aminotransferase since it is easy to show that neither entry of 2-oxoglutarate nor the action of malate dehydrogenase is rate limiting (Marra *et al.*, 1977b).

The essentials of measurement of uptake of enzyme into mitochondria are then as follows. The endogenous aspartate aminotransferase activity is measured in a sample of mitochondria as described above. To a second sample is added purified isoenzyme before addition of 2-oxoglutarate; the increase in the rate of change of fluorescence compared with the control is then a measure of enzyme uptake into the organelles. A possible objection to this interpretation is that some enzyme and substrates remain outside the organelles and hence the observed effect on the rate of change of fluorescence may be due to some extramitochondrial effect. This objection is conclusively overcome by the observations that if cytosolic aspartate aminotransferase is used instead of the mitochondrial form, then no increase in rate of change of fluorescence is observed. Hence in our system the uptake of mitochondrial aspartate aminotransferase is specific in that the cytosolic isoenzyme is excluded.

We have also demonstrated the same phenomenon by a different method. Asparate aminotransferase can be radiolabeled by reduction of the Schiff base linking the cofactor to the protein with sodium borotritiide. It is then possible to measure uptake of the enzyme into mitochondria by incubation of enzyme with the organelles, centrifugation, and counting of the radioactivity incorporated. Again, labeled cytosolic isoenzyme may be used as a control. Using this method, we were able to show (Marra *et al.*, 1978) that mitochondrial aspartate aminotransferase can interact in two different ways with mitochondria. At high ionic strength, only uptake into the organelles occurs; this is shown for example by the fact that the sequestered enzyme is rendered insensitive to protease added externally (Marra *et al.*, 1980a). At low ionic strength there is a combined effect of uptake into the organelles and binding externally; these effects can be differentiated since externally bound enzyme is highly protease sensitive. Another difference between the two phenomena is that measurement of uptake of isoenzyme as a function of amount added externally to the organelles shows a hyperbolic increase followed by saturation whereas external binding shows a linear increase and does not saturate in the concentration range used (Marra *et al.*, 1978). This external binding may be expected since the enzyme is basic and indeed binds to several different membrane systems at low ionic strength (Marra *et al.*, 1977a) as well as to negative liposomes (Furuya *et al.*, 1979). Recent evidence suggests however (Section VI,D) that this binding may not be due solely to nonspecific electrostatic effects.

In summary, both of the methods described can be used to measure uptake of mitochondrial aspartate aminotransferase into mitochondria. The fluorescence method is the procedure of choice since it measures only enzyme which has reached the matrix of the mitochondrion; the assay method is such as to detect only enzyme which has reached this compartment. The method using labeled isoenzyme could, in principle, measure enzyme sequestred in some other location inside the mitochondria although we have no reason to believe that this is so.

In order to assess the generality of the model system described above we subsequently extended the work to the isoenzymes of malate dehydrogenase (Passarella *et al.*, 1980, 1983). A fluorescence method was again used to measure intramitochondrial malate dehydrogenase activity. The method involves transport into the organelles of oxaloacetate; malate dehydrogenase activity then results in oxidation of NADH and concomitant decrease in fluorescence due to this substance. In the case of rat liver mitochondria, however, because of the low rate of uptake of oxaloacetate and the high endogenous enzyme activity, oxaloacetate uptake is rate limiting. This can be overcome by using mitochondria loaded with sulfite in which endogenous enzyme activity is partially inhibited. The rate of change of fluorescence of NADH then gives a true measure of residual enzyme activity. Under these conditions, incubation of the organelles with purified mitochondrial malate dehydrogenase leads to an increased intramitochondrial enzyme activity i.e., the enzyme is taken up into the mitochondria. As with aspartate aminotransferase, the cytosolic isoenzyme was used as a control and did not produce an increase in intramitochondrial enzyme activity; that is, uptake is again a specific process. The need to use mitochondria loaded with sulfite for experiments with malate dehydrogenase may be considered a disadvantage. In this context it is of interest that rat kidney mitochondria, in which endogenous enzyme is less active and oxaloacetate uptake more rapid, can be used for measurement of enzyme import without pretreatment with sulfite (S. Passarella, unpublished work). This last observation is of interest in that it suggests that the protein uptake machinery is the same or very similar in mitochondria from different tissues of the same animal species. As an extension to this it has been shown using *in vitro* systems (Mori *et al.*, 1982; Freeman *et al.*, 1983) that uptake of newly synthesized proteins into mitochondria from other animal species also occurs.

Our claim that native mitochondrial proteins can be translocated across the mitochondrial membrane system is not unique. Waksman and co-workers (Waksman and Rendon, 1974; Waksman *et al.*, 1977) claim that in response to "movement effectors" native proteins can be translocated reversibly from the mitochondrial matrix into the intermembranal space (reviewed by Waksman *et al.*, 1980). For example, it is claimed that for mitochondria incubated in sucrose alone the distribution of aspartate aminotransferase is 9% intermembranal, 79% innermembrane-bound, and 12% in the matrix, whereas in the presence of fumarate these values become 79, 6, and 15%, respectively; other compounds result in other distributions. On removal of the "movement effector" these translocations are reversed and the original distributions reestablished. These results were obtained by fractionation of the mitochondria and determination of the distribution of enzyme activity in each fraction; this is difficult to do without causing redistribution of the enzyme between the fractions.

In a somewhat different experimental system (Hubert *et al.*, 1979) binding and internalization of mature aspartate aminotransferase by mitoplasts and submitochondrial particles were claimed. The experimental techniques were based on measurements of latency of enzyme activity, protease sensitivity, and antibody binding, and seemed to be much less susceptible to artifacts than those used in the earlier work. Hence it seems that such protein movements do occur but their physiological significance is not clear. It should be emphasized that the model system used by us is not the same as that of Waksman and co-workers. An obvious difference is that in our system transport of proteins is unidirectional. Nevertheless, the results outlined above give added weight to the idea that mature proteins can move through mitochondrial membranes.

VI. Results of Studies of Transport of Proteins into Mitochondria

For clarity, the discussion below is organized into topics of interest rather than into a description of work with individual enzymes. An attempt is made to synthesize results obtained with one or more of the experimental approaches described in Section V.

A. Proteolytic Processing of Precursors

As outlined in Sections V,A and V,B, import of precursors carrying N-terminal extensions into mitochondria is associated with processing to the mature size.

A protease involved in processing has been isolated from yeast mitochondria and appears to be localized in the matrix (Böhni *et al.*, 1980; McAda and Douglas, 1982; Schatz and Butow, 1983). The protease is metal dependent and is sensitive to chelators. It is very specific, acting only on precursors of mitochondrial proteins from which it generates the correct N-terminus of the mature protein; it does not act on mature mitochondrial proteins nor on nonmitochondrial proteins. It is of interest that this protease acts on the precursor of subunit V of cytochrome *c* oxidase, an integral membrane protein. If it is generally involved in processing inner membrane proteins then this requires that in all such cases the N-terminal prepiece is inserted completely through the mitochondrial inner membrane, since, as stated above, the enzyme is found in the matrix.

Until recently there was no information available on the timing of proteolysis of precursors. Zwizinski and Neupert (1983) have now shown, however, that the precursors of the β-subunit of F_1-ATPase and of subunit 1X of the F_0-ATPase are imported into *Neurospora* mitochondria in the absence of proteolytic processing. The very clear distinction in this regard between uptake of proteins into mitochondria and the processes of protein export from cells, where cleavage of the signal peptide occurs at a very early stage, is noteworthy.

Surprisingly, uptake of at least some proteins into the intermembranal space and their processing involve this same matrix protease. It has been shown that the precursors of both cytochrome b_2 and cytochrome c_1 are proteolyzed in two steps, the first step generating intermediates of sizes between those of the precursors and the mature forms (Gasser *et al.*, 1982a); cytochrome c_1 is not properly an enzyme of the intermembranal space since part of the polypeptide chain is inserted into the inner membrane, but nevertheless its processing is similar to that of cytochrome b_2. Similar behavior is also shown by cytochrome *c* peroxidase (Reid *et al.*, 1982). It is proposed for all of these enzymes (Gasser *et al.*, 1982b; Reid *et al.*, 1982; Daum *et al.*, 1982; Ohashi *et al.*, 1982) that the N-terminus of the precursor passes through the mitochondrial membrane in an energy-dependent step (see Section VI,B) and is cleaved by the matrix protease. A second proteolytic event catalyzed by an enzyme possibly on the outer face of the inner membrane then generates the protein of mature size. In the case of cytochrome c_1 the second proteolytic cleavage is thought to be preceded by addition of heme with final proteolysis occurring as a slow last step in formation of the mature protein (Ohashi *et al.*, 1982).

The sequence of events proposed in the case of cytochrome *c* peroxidase correlates well with the features of the N-terminal sequence of the pre-protein as determined from DNA sequence analysis by Kaput *et al.* (1982). These authors find an 18-residue highly charged sequence at the N-terminal end, followed by a sequence of 23 residues of a very nopolar nature which is predicted to be α-helical, followed by another relatively polar sequence of 27 residues before the N-terminus of the mature protein. It is proposed that the first of these sequences crosses the inner membrane into the matrix whereas the central apolar region constitutes a ''stop-transfer'' signal leaving the remaining part of the presequence in the intermembranal space. Cleavage of the appropriate peptide bonds then generates the mature protein.

Knowledge of the location and properties of the processing activity is not restricted to yeast. Mori and his colleagues have purified from rat liver mitochondria an enzyme which accurately processes ornithine transcarbamylase (Miura *et al.*, 1982a). This enzyme is inhibited by chelators and also by inhibitors of serine proteases. It is specific in that it did not cleave precursors of secretory proteins, but interestingly it also did not process the precursor of carbamoyl-phosphate synthetase; this suggests that liver mitochondria contain more than one processing enzyme. This work has been reviewed by Mori *et al.* (1982). It should also be mentioned that ornithine transcarbamylase appears to be processed in two steps (Morita *et al.*, 1982). The significance of this is not clear.

Recently Ono and Ito (1982) have studied the processing of the precursor of sulfite oxidase (an intermembranal enzyme) by rat liver mitochondria. Fractionation of mitochondria suggested that the processing activity was associated with the outer face of the inner membrane. Hence it may be that the mechanism of processing of this protein is not the same as with the other intermembranal proteins described above.

B. Energy Dependence of the Uptake Process

Mention has been made in Section V,B that both in yeast cells and in animal cells the presence of uncouplers of oxidative phosphorylation leads to the accumulation of precursors of mitochondrial proteins in the cytosol. One of the first demonstrations that some aspect of protein import into mitochondria is energy dependent came from the work of Nelson and Schatz (1979) on processing of precursors of the three large subunits of F_1-ATPase and of the two subunits of the cytochrome bc_1 complex. The suggestion was made that either the processing or the transfer of precursors across the mitochondrial membrane was ATP dependent. Similar observations were made by other groups. For example Mori *et al.* (1981b) showed that processing of preornithine transcarbamylase *in vitro* was inhibited by dinitrophenol and FCCP but not by KCN or NaN_3; the possibility was recognized by these authors that the inhibitors might interact with receptors for protein import, with the protease required for processing, or some other component required for protein import. Waksman *et al.* (1980) have made the same point. Lewin *et al.* (1980) using yeast mitochondria depleted of ATP or using CCCP as an inhibitor showed failure of yeast cells to process the precursors of subunits V and VI of cytochrome *c* oxidase. Similarly, Autor (1982) showed that deenergized mitochondria from yeast failed to process the precursor of Mn-superoxide dimutase. Also Zimmermann and Neupert (1980) reported that transport of subunit IX of F_0-ATPase and of the ADP/ATP translocator into *Neurospora* mitochondria was inhibited by FCCP. Conversely there was no effect on uptake of apocytochrome *c* (Zimmermann *et al.*, 1981). This last result, however, does not exclude energy requirement for the uptake of apocytochrome *c*, but only implies that this protein does not use for its processing and proper insertion into the outer face of the inner membrane either ATP or a transmembrane potential.

None of the above-cited reports provided sufficient information to distinguish between a requirement for ATP per se in the transport of proteins into mitochondria or for a transmembrane electrochemical gradient. Distinction between these possibilities requires a comparative study of the effect of uncouplers, respiratory chain inhibitors, and ionophores. Particularly in the case of ionophores, control of the ionic environment is essential for interpretation of results; this control is difficult with intact cells or in the systems described in Section V,A for studying uptake of precursors into mitochondria *in vitro*. It is however relatively straightforward using our model systems as described in Section V,C. Using these model systems we have shown (Passarella *et al.*, 1982a, 1983) that uptake of native aspartate aminotransferase and malate dehydrogenase into mitochondria depends on a transmembrane electrochemical gradient. Moreover in both cases it is the ΔpH component rather than Δψ (using the terminology of Mitchell, 1968) which is important. The essential points which demonstrate this fact are as follows taking aspartate aminotransferase as an example (Passarella *et al.*, 1982a). Uptake into mitochondria is inhibited by antimycin and by cyanide as well as by FCCP (Marra *et al.*, 1980b). On the other hand uptake is increased by ATP. These results could be taken to show a direct involvement of ATP or of a transmembrane potential. However, the uptake process is stimulated by oligomycin, an inhibitor of ATP synthesis, which argues strongly for the involvement of a transmembrane ion gradient. Confirmation of this and demonstration that ΔpH is the essential component was provided by results obtained using ionophores. Ionophores of the valinomycin class in media containing high K^+ concentration decrease Δψ but increase ΔpH. Uptake of aspartate aminotransferase into mitochondria was strongly enhanced by valinomycin; the addition of phosphate ions collapses ΔpH in the presence of valinomycin and consistently uptake was abolished under these conditions. In the case of nonactin, the distribution of K^+ and H^+ depends on the concentration of external K^+ (Palmieri and Quagliariello, 1969). In the presence of 10 mM K^+, where ΔpH is largely unaltered, no effect on the uptake of the enzyme was observed. At higher concentration of K^+, ΔpH is increased as was uptake of the enzyme. Gramicidin, which increases ΔpH but decreases Δψ stimulated the uptake process. Finally, and most significantly, nigericin which promotes an electroneutral exchange of K^+ and H^+ thereby collapsing ΔpH, while maintaining Δψ, completely inhibits uptake of the enzyme into mitochondria. As a final demonstration of the importance of a transmembrane pH gradient, it was shown that as the pH of the incubation medium was decreased, thus increasing ΔpH, enzyme uptake increased.

Essentially identical results were obtained with malate dehydrogenase (Passarella et al., 1983). In addition, use was made of safranine, a substance whose absorption is sensitive to variation of $\Delta\psi$ across membranes. Addition of malate dehydrogenase and its uptake into mitochondria treated with safranine caused only a slight change in absorbance of the safranine, consistent with a marginal increase in $\Delta\psi$ and decrease of ΔpH. Again these results are consistent with a dependence on ΔpH and not on $\Delta\psi$ for uptake of malate dehydrogenase into mitochondria.

Subsequently reports have appeared from other laboratories demonstrating the requirement for an electrochemical potential for protein import into mitochondria. Gasser et al. (1982b), for example, showed that import of precursors by isolated mitochondria was blocked by protonophores or by valinomycin plus K^+, but not by oligomycin. Under the conditions used it was not possible to distinguish, as the authors acknowledged, between a dependence on ΔpH or on $\Delta\psi$. Similar results have been reported by Mori et al. (1982) for uptake of ornithine transcarbamylase into rat liver mitochondria.

An attempt to distinguish between the two components of the transmembrane electrochemical gradient as the energy source for protein uptake has been made by Schleyer et al. (1982) in the case of transport into Neurospora mitochondria of the precursors of subunit IX of the F_0-ATPase and of the ADP/ATP translocator. It was concluded that it is the $\Delta\psi$ component of the membrane electrochemical gradient which is involved in the process. The main lines of evidence employed were that the process was blocked by valinomycin at high K^+ concentration, but not by nigericin, in contrast with the result reported by us (Passarella et al., 1982a, 1983). Schleyer et al. (1982) suggest that this discrepancy arises from the fact that transfer into mitochondria of precursors and of mature proteins are different processes. We would maintain, however, that the different results obtained may be due to the somewhat uncontrolled conditions used by Schleyer and co-workers. To measure protein import in their system (as well as in others, see Section VI,C), incubation of mitochondria with precursors for extended periods (1 hour) at 25°C is required and it seems unlikely that mitochondrial integrity would be maintained under these conditions; no attempt was made to show that the mitochondria retained their ability to respire or maintain transmembrane gradients (as was done by Passarella et al., 1983) which should be considered as an essential control for experiments of this type. Second the effects of ionophores on ΔpH and $\Delta\psi$ are very dependent on K^+ concentration, a parameter which was not carefully controlled in the work reported. Hence we consider that the results of Schleyer et al. (1982) need confirmation under carefully controlled conditions and that the weight of experimental evidence at this time supports the involvement of the transmembrane pH gradient as the driving force for uptake of proteins into mitochondria.

Considerable support for this view has recently been provide by the work of Kitagawa *et al.* (1984). These authors studied the uptake of the precursor of carbamoyl-phosphate synthetase into mitochondria in cultures of Reuber hepatoma H-35 cells. Uptake was inhibited by low concentrations (0.5 μM) of nigericin and monensin. The effects of valinomycin and nonactin could not be established at the same concentration because protein synthesis was inhibited; at lower concentrations there was no effect by these compounds. The similarity between the observed effects of nigericin on protein uptake *in vivo* and in our model system (see above) is striking.

C. Rate of Import of Proteins into Mitochondria

Although the rate of uptake and processing of precursors by isolated mitochondria *in vitro* (see Section V,A) has not been explicitly mentioned in the many publications describing work of this type, it is obviously a slow process. Typically, incubation times of 30 minutes to 1 hour have been used and uptake is frequently not complete in this time (see for example, Harmey *et al.*, 1977; Maccecchini *et al.*, 1979a; Mori *et al.*, 1982; Gasser *et al.*, 1982b). This has led to the view that the uptake process *in vivo* is also slow. In contrast import of mature proteins in the model systems described in Section V,C is very rapid being essentially complete in 1 minute; this has caused some doubts as to the validity of model systems as a reflection of the physiological process.

Recently measurements of the rate of import of precursors into mitochondria in whole cells have been made and it is clear that the process *in vivo* is indeed very rapid. Jaussi *et al.* (1982) have studied the synsthesis of mitochondrial aspartate aminotransferase in chicken embryo fibroblasts by pulse labeling; in a subsequent chase the precursor was converted into mature enzyme with a half-life of 30–60 seconds. Similarly, in isolated rat hepatocytes during a chase after pulse labeling, precursors of ornithine transcarbamylase and carbamoylphosphate synthetase were imported into mitochondria with half-lives of 1–2 minutes (Mori *et al.*, 1981a; Miura *et al.*, 1982b). A similar rate was observed for uptake of the precursor of carbamoyl-phosphate synthethase in rat liver explants (Raymond and Shore, 1981). This rapidity of uptake is not restricted to animal cells nor to matrix proteins. Reid and Schatz (1982b) have shown that in yeast, uptake of the β-subunit of F_1-ATPase has a half-life of less than 1 minute. Similarly Ades and Butow (1980b) have obtained results suggesting a half-life of uptake of proteins into yeast mitochondria *in vivo* of about 1.5 minutes.

A somewhat greater half-life was observed for formation of holocytochrome *c* in *Neurospora* cells (Hennig and Neupert, 1981) but this may be due to a rate-limiting addition of heme rather than of import of the precursor. The biosynthesis and transport of cytochrome *c* peroxidase is interesting in this context (Reid *et al.*, 1982). Formation of the mature protein is relatively slow, but it appears that the precursor is translocated into mitochondria and the presequence inserted through the inner membrane very rapidly (half-life much less than 3 minutes); subsequent proteolysis to yield the mature protein is the rate-limiting step.

The picture that emerges, therefore, is of extremely rapid uptake of precursors into mitochondria *in vivo*. Apart from the inherent interest of this high rate of uptake, it also raises the question why uptake of precursors into mitochondria *in vitro* is so slow in systems such as those described in Section V,A. Clearly those systems do not reflect accurately the process occurring *in vivo* and lead one to suppose that, during the prolonged incubation periods used, the mitochondria lose their integrity and hence their ability to take up precursors rapidly; this possibility was alluded to in a different context in the previous section.

On the other hand, the rapidity of uptake *in vivo* provides strong support for the view that the model systems that we have developed (see Section V,C) reflect accurately the events occurring *in vivo*. Recently we have carried out a detailed investigation of the kinetics of uptake of both native aspartate aminotransferase and native malate dehydrogenase into mitochondria. In both cases the half-lives obtained were of the order of 0.25 minute (Marra *et al.*, 1982a; and authors' unpublished work). These kinetic studies will be discussed in more detail in Section VI,E.

There is another interesting aspect to some of the work described above related to the fate of mitochondrial precursor proteins whose import into mitochondria is blocked; several of the studies were performed by preventing import by addition of uncoupler or other means and then observing import when the constraint was removed. In some cases, precursor whose import into mitochondria or processing was blocked was rapidly degraded. For example inhibition of the protease responsible for processing led to a rapid degradation of pre-carbamoyl-phosphate synthetase in liver explants with a half-life of 2–3 minutes (Raymond and Shore, 1981). Similarly, in chicken embryo fibroblasts blocked for protein uptake by CCCP the precursor of aspartate aminotransferase was degraded with a half-life of about 5 minutes (Jaussi *et al.*, 1982). In yeast, the rate of degradation of precursors blocked for uptake into mitochondria varies. The precursor of apocytochrome c_1 was degraded with a half-life of about 10 minutes whereas the precursor for the β-subunit of F_1-ATPase was very stable and accumulated to high levels in the cells (Reid and Schatz, 1982a). The significance of these observations and the site of degradation of unimported precursors are by no means clear.

D. CHEMICAL MODIFICATION OF PROTEINS AND ITS EFFECT ON IMPORT INTO MITOCHONDRIA

One obvious way in which to obtain information about the mechanistic aspects of transport of proteins through mitochondrial membranes is by carrying out chemical modifications and observing the effect on the import process. Clearly this is not possible in systems using intact cells (Section V,B). Similarly, to carry out this sort of work with precursors of mitochondrial proteins would require their isolation in quantities sufficient for chemical manipulations: this has not yet been done, but the accumulation of some precursors in yeast cells blocked for protein import offers the possibility of their isolation and chemical modification.

Up to now, the only system which has allowed any work on the effect of chemical modification on uptake is the model system described in Section V,C involving uptake of mature aspartate aminotransferase and malate dehydrogenase into mitochondria. The former protein is a particularly useful object of study in this context since its complete amino acid sequence is known, as is that of the cytosolic isoenzyme from several sources (Christen *et al.*, 1984).

A feature of mitochondrial aspartate aminotransferases that has been extensively studied is their possession of a single highly conserved sulfhydryl group exposed in the native enzyme; other sulfhydryl groups are not accessible to chemical reagents under native conditions (see Petrilli *et al.*, 1981, for a review of these studies). This prompted us to examine the possibility that this sulfhydryl group is involved in the uptake of the enzyme into mitochondria (Marra *et al.*, 1979). It was shown that reaction of this single sulfhydryl group in rat liver mitochondrial aspartate aminotransferase with reagents such as mersalyl abolished uptake of the enzyme into mitochondria. Significantly treatment of the mitochondria themselves with mersalyl did not affect their ability to internalize the enzyme. In the case of malate dehydrogenase (Passarella *et al.*, 1983) a similar effect was observed but it was not possible to establish whether blockage of one or a small number of sulfhydryl groups was responsible for inhibition of transport of the enzyme. These results, as well as showing that very minor chemical modifications inhibit protein uptake into mitochondria, also led to experiments designed to establish the existence of a receptor for those proteins in the mitochondrial membrane system (see Section VI,E).

Another structural feature which is very highly conserved in mitochondrial aspartate aminotransferase is the N-terminal amino acid sequence; the first 20 amino acids are identical, with one exception, in all the mitochondrial isoenzymes studied to date whereas the cytosolic isoenzymes are quite variable in this region (Bossa *et al.*, 1981). It has already been shown (Sandmeier and Christen, 1980) that the N-terminal segment of aspartate aminotransferase from chicken can be removed proteolytically with little alteration to the rest of the structure. We have now shown that this can be also done with the enzyme from rat liver and moreover that the proteolysed enzyme obtained is not imported into mitochondria (O'Donovan *et al.*, 1984). It is also of interest that the protein lacking the N-terminal segment does not bind to the outer face of mitochondria as does the mature form (see Section V,C). This suggests that the binding phenomenon has some physiological significance and is not a nonspecific electrostatic effect; the protein remaining after removal of the N-terminal section is as basic as the native enzyme. Taken together, these results suggest that the N-terminal segment of native aspartate aminotransferase is of central importance in transport of the protein into mitochondria. This is reminiscent of the situation for export of ovalbumin where the signal sequence for export has been shown to be in the N-terminal section of the mature protein (Meek *et al.*, 1982).

An alternative approach to study of the effects of structural change on uptake of proteins into mitochondria is provided by recent success in cloning the genes for some mitochondrial proteins (see Section IV,C). For example, site directed mutagenesis could be used to selectively change individual residues in the proteins or recombinant DNA techniques could be used to make more extensive changes. An example of this sort of approach, where the modified gene was obtained fortuitously, is provided by the work of Riezman *et al.* (1983c). Two forms of the gene for a protein of the outer mitochondrial membrane were isolated from a genomic clone bank, one of the intact gene and the other a truncated version lacking a large part of the 3'-terminus. Both genes were transformed into yeast and it was shown that the truncated version led to expression of a protein lacking 203 amino acid residues at the C-terminus. In spite of the extensive deletion, this shortened protein still inserted into the mitochondrial surface, thus showing that the addressing and insertion signals for this protein are located in the N-terminal section, the C-terminus of the protein being dispensable for these functions (but not for biological activity). The membrane anchoring domain may be provided by a sequence of 28 uncharged amino acids comprising residues 10 to 37 (Hase *et al.*, 1983). It is likely that approaches of this sort will produce much interesting information in the future.

E. Evidence for the Existence of Receptors for Protein Transport into Mitochondria

Evidence is now accumulating for the presence of specific receptor sites for import of mitochondrial proteins into mitochondria. Early results were those of Zimmermann and Neupert (1980) with the ADP/ATP translocator protein. It was shown that in cell-free systems the precursor form of the protein bound rapidly to added mitochondria, but remained protease sensitive; only at a later stage was it integrated into the inner membrane with concomitant loss of protease sensitivity. More recently Zwizinski *et al.* (1983) have shown that, in the absence of a transmembrane potential, the precursor protein binds to mitochondria but is not imported. Upon reestablishment of the potential, the bound precursor is imported without dissociation from the mitochondria suggesting that the external binding is a step on the import pathway.

Similar work with apocytochrome *c* (Hennig and Neupert, 1981) made use of the fact that import of this protein to the outer face of the inner mitochondrial membrane is inhibited by hemin. In the presence of this substance, substantial binding of newly synthetized apocytochrome *c* to the outer membrane in a protease sensitive fashion was observed. The binding showed the characteristics of interaction with a specific receptor in that it was saturable, rapid, and reversible by addition of excess unlabeled apocytochrome c. More recently, Hennig *et al.* (1983) have measured the number of receptor sites in *Neurospora* mitochondria and their affinity for apocytochrome c. The functional equivalence between newly synthesized and chemically prepared apocytocrome c was exploited in experiments in which it was shown that a large excess of the latter did not inhibit uptake into mitochondria *in vitro* of the precursor of the ADP/ATP translocator or of the dicyclohexylcarbodiimide binding protein of ATP synthase; uptake of cytochrome c was inhibited (Zimmerman *et al.*, 1981). This was taken to show that the receptors for cytochrome c and for the other two proteins are distinct. Miura *et al.* (1982b) have also pubished some observations relevant to this subject. They showed that the precursor of ornithine transcarbamylase is more basic than the mature protein. Uptake of this precursor into mitochondria was inhibited by basic proteins such as histones and protamin but not by acidic proteins. This could be taken to indicate that the positively charged prepiece interacts with a negatively charged region of the mitochondrial membrane system and that this interaction is competitively inhibited by positively charged proteins.

The results discussed above indicate that receptor sites exist for the proteins mentioned, but reveal little about the nature of these receptors. We have used two approaches to this problem. One arose out of the observations quoted in Section VI,D that reaction of sulfhydryl groups with reagents such as mersalyl blocks uptake of aspartate aminotransferase and malate dehydrogenase into mitochondria. This suggested to us the possibility that uptake involved interaction of protein thiols with specific binding sites in membrane receptors. This possibility was tested by examination of the effects of 2-mercaptoethanol added to the systems used to study protein import (Passarella *et al.*, 1982b, and authors' unpublished work). Uptake of both isoenzymes was inhibited and inhibition was shown to be due to interaction of 2-mercaptoethanol with the mitochondria and not with the isoenzymes themselves. Inhibition was relieved by *N*-ethylmaleimide and mersalyl. Interestingly, the response of the two isoenzymes to inhibition of uptake by 2-mercaptoethanol was different. Uptake of aspartate aminotransferase was inhibited competitively; that of malate dehydrogenase was less sensitive to 2-mercaptoethanol and the inhibition was of the mixed type. These results strongly suggest that the binding sites for the two enzymes are provided by proteins rather than lipids and moreover that the receptors are not the same. Further evidence for this was provided by measurements of the kinetics of uptake of the two isoenzymes and by their mutual inhibition (Marra *et al.*, 1982a, and authors' unpublished work). The inhibition of uptake of one isoenzyme exerted by the other was shown to operate at the level of the membrane system (rather than by competitor that had already been taken up). Kinetic analysis further showed that the inhibition exerted by one isoenzyme on the uptake of the other was, at least approximately, uncompetitive. This type of inhibition can only be rationalized by a model in which the two isoenzymes have different but interacting receptors. Binding of one isoenzyme to its receptor is envisaged to increase the affinity of the neighboring receptor for the other isoenzyme, perhaps by a conformational change, but the modified receptor is not competent in protein import. Binding of the inhibitor to this modified receptor similarly inactivates the receptor already occupied. This type of mechanism yields a mixed inhibition but examination of the rate equations showed that the kinetics approximate to uncompetitive with the values of the binding constant obtained experimentally.

The question of which of the mitochondrial membranes contains the putative binding sites remains to some extent open. We have shown that mitoplasts lacking the external membrane internalize aspartate aminotransferase and malate dehydrogenase in a manner which is indistinguishable from that found with intact mitochondria (Marra *et al.*, 1982b). These results tend to implicate the inner membrane as the site of the receptors. It should be recalled, however, that Butow and his colleagues have shown that binding of ribosomes thought to be active in protein import in yeast mitochondria occurs at positions where the inner and outer membranes are in close contact (Schatz and Butow, 1983). Hence the possibility arises that protein import into mitochondria occurs at points where the two membranes function as a single unit (Chua and Schmidt, 1979) or indeed there may be regions where the two membranes fuse, perhaps transiently, to form a single physical unit as the site of protein import.

Against this view, Schatz and his colleagues have recently claimed that import receptor sites are in the outer membrane of yeast mitochondria. Outside-out vesicles formed from the outer membrane of yeast mitochondria were isolated and characterized (Riezman *et al.*, 1983a). It was then shown (Riezman *et al.*, 1983b) that these vesicles bind the precursors of cytochrome b_2 and of citrate synthase but not the partially processed precursor of cytochrome b_2 nor non-mitochondrial proteins. Very little binding of pre-cytochrome b_2 to the inner membrane was observed; on the other hand, selectivity was much less with pre-citrate synthase. The authors take these results to show that binding to a receptor in the outer membrane is an obligatory step in import of proteins of the matrix and intermembranal space.

Whatever the nature and location of the receptors, recent results show that the recognition of mitochondrial proteins, and indeed the processes of import and processing, have been very highly conserved during evolution. For example Takiguchi *et al.* (1983) have reported that rat liver ornithine transcarbamylase is taken up and processed by mitochondria from pigeon liver, frog liver, and carp liver (but not by mitochondria from yeast). Even more striking is the observation (Schmidt *et al.*, 1983b) that yeast mitochondria take up and process the precursor of *Neurospora* ATPase subunit IX even though the corresponding protein in yeast is a mitochondrial gene product and hence is inserted into the inner membrane from the matrix side. Other examples of uptake of proteins into mitochondria in heterologous systems were outlined in Section V,C.

VII. Speculations on the Mechanism of Protein Transport into Mitochondria

A. DIVERSITY OF UPTAKE MECHANISMS

This final section will conclude with a review of the results given in the previous sections and an attempt to synthesize the various observations into a more coherent picture.

First, it must be acknowledged that, given the diversity of locations for imported proteins, it is very unlikely that there will be a unique mechanism for import of all proteins. Indeed, evidence for diversity of mechanisms is already available. In the case of porin, for example, the protein is synthesized with the native molecular weight, and its incorporation into the outer membrane does not depend on energization of the inner membrane (Freitag et al., 1982; Mihara et al., 1982b). Hence it seems likely that this protein spontaneously inserts into the outer membrane and mechanisms such as those proposed by Wickner (1979, 1980, 1983) or by Engelman and Steitz (1981) may operate in this case. The picture for at least some intermembranal proteins is obviously very different as shown by Schatz and co-workers. Available evidence suggests that the presequence of proteins such as cytochrome c peroxidase, cytochrome b_2, and cytochrome c_1 (properly an inner membrane protein) pass through the mitochondrial inner membrane leaving the rest of the protein outside (Daum et al., 1982; Ohashi et al., 1982; Reid et al., 1982). From the amino acid sequence of the prepiece of cytochrome c peroxidase it seems likely that the polypeptide passes through the membrane as a single helical chain, rather than as a double helical hair-pin as in the model of Engelman and Steitz (1981). The reason for this partial excursion into the inner membrane is not clear, but an attractive hypothesis might be that interaction occurs at points where the inner and outer membranes have fused and that the penetration of the presequence serves to anchor the precursor until a change in outer membrane encloses the protein; proteolysis at the outer face of the inner membrane would then liberate the mature protein into the intermembranal space (Daum et al., 1982). There are obvious differences also in the detailed mechanism of biogenesis and topogenesis of different proteins of the inner membrane as well as between proteins destined for different compartments. Comparing cytochrome c and the ADP/ATP translocator for example, although neither of these proteins is synthesized as a precursor of molecular weight greater than the mature form in *Neurospora crassa* (Hennig and Neupert, 1981; Zimmermann et al., 1979) in contrast with the majority of inner membrane proteins (Table I), the similarity between their mechanisms of uptake seems to be limited. They differ fundamentally in that translocation of the former to its site of action does not require an energized inner membrane whereas energization is essential for topogenesis of the translocator (Schleyer *et al.*, 1982). Moreover, the two proteins use different receptors for their import (Zimmermann *et al.*, 1981). It is not yet known whether either of these receptors is shared by other inner membrane proteins or by proteins of the matrix and intermembranal space.

Finally it should always be borne in mind that there may be differences between species in relation to mechanisms of mitochondrial biogenesis. A very obvious case in point relates to subunit IX of the F_0-ATPase which is coded by a mitochondrial gene in yeast (Macino and Tzagoloff, 1979; Hensgens *et al.*, 1979) but by a nuclear gene in *Neurospora* (Sebald, 1977; Michel *et al.*, 1979). A more subtle difference is evident for the ADP/ATP translocator protein which is synthesized as a precursor of higher molecular weight than the native form in rat liver (Chien and Freeman, 1983) but not in *Neurospora* (see above). Other species differences may become apparent when more information is obtained.

In summary, then, it would probably be mistaken to attempt to formulate a single mechanism for protein uptake into mitochondria since such a mechanism in unlikely to exist. What follows, therefore, are comments on various interesting features of the uptake process which may not apply to proteins destined for all mitochondrial compartments nor necessarily for the same protein from different species.

B. The Role of the Prepiece

The majority of inner membrane and matrix proteins are synthesized as precursors of molecular weight greater than the mature forms (Table I). It is therefore, necessary to examine what the role of these prepieces may be. Views on this have been colored by the obvious analogy with secretory proteins where there seems little doubt that N-terminal extensions play a critical role in transport across the endoplasmic reticulum (Blobel and Dobberstein, 1975). It must be recalled, however, that protein transport through mitochondrial membranes is fundamentally different from protein export in several respects and in particular from the point of view that the former is posttranslational and the latter cotranslational. In addition, there is no direct evidence that the prepieces of inner membrane or matrix proteins function as leader sequences in the sense of directing proteins into or through the inner membrane (but see above in relation to proteins of the intermembranal space). Indeed the striking different characteristics reported for the presequences of aspartate aminotransferase, which has an acidic presequence (Kamisaki *et al.*, 1982), and for ornithine transcarbamylase, which has a basic presequence (Miura *et al.*, 1982b), argues against such a unique function.

On the other hand, there is some evidence to show that presequences are not necessary for protein transport. The most extensive evidence for this comes from our own work on uptake of mature proteins into mitochondria (see Section V,C). Although the validity of these observations has been questioned (see for example, Ades, 1982) we consider that there is no room for doubt that our model system measures uptake of mature proteins into the mitochondrial matrix. Indeed, after incubation of mitochondria with radiolabeled aspartate aminotransferase, removal of externally bound enzyme by proteolysis (Marra *et al.*, 1980a) and rupture of the organelles, the presence of radiolabeled enzyme has been demonstrated on SDS–polyacrylamide gel electrophoresis followed by autoradiography (K. M. C. O'Donovan and S. Doonan, unpublished work).

There have been isolated reports that mature proteins do not compete with precursor proteins for uptake into mitochondria in *in vitro* systems (see, for example, Sakakibara *et al.*, 1981, for aspartate aminotransferase). It should be recalled however, that precursor processing in those reconstituted systems is extremely slow compared with the situation *in vivo* and also in our model system. Hence a rapid uptake of native protein followed by a slow uptake of precursor need not produce an obvious inhibition of the latter. In addition, mitochondria of high quality are required in our model system, whereas the slow uptake of precursors into mitochondria *in vitro* frequently observed calls into doubt the integrity of the mitochondria used.

The results quoted in Section IV,D are also of interest in this context. In at least three cases proteins which are enzymatically active in the cell cytosol and hence must have native structures, nevertheless can be taken up into the mitochondria. In the cases of fumarase and malonyl-CoA carboxylase the cytosolic isoenzyme plays a physiological role. Although the uptake of these proteins seems to involve proteolysis, albeit of a very small piece of polypeptide in the case of fumarase, this does not seem to be analogous to the precursor–product relationship found for other mitochondrial proteins. More recently it has been shown (Felipo *et al.*, 1983) that the precursor of glutamate dehydrogenase has enzymic activity. Hence in this case also the precursor must have a conformation which is very close to that of the mature enzyme.

Given that native mature enzymes can be imported into mitochondria *in vitro* this suggests that the prepieces of precursor proteins must, in most cases, play some other role or roles than that of leading proteins into or through the mitochondrial membranes. What might these roles be?

One obvious possibility is that presequences act as addressing signals *in vivo* and direct the newly synthesized protein to the mitochondria, insertion into the membrane system or passage through it being mediated by other parts of the structure of the protein. It is of interest in this context that Anderson *et al.* (1983) have provided clear evidence for a distinction between addressing signals and insertion signals in the case of integration of proteins into microsomal membranes. Integration of three proteins into dog pancreas microsomal membranes was studied, namely calcium ATPase of rabbit sarcoplasmic reticulum, protein MP26 of bovine lens fiber plasma membrane, and rat liver cytochrome b_5. These proteins are similar in that none of them is synthesized with an N-terminal extension. They were different, however, in that the first two required signal-recognition particle for integration whereas the cytochrome b_5 did not. It was concluded that cytochrome b_5 contains an insertion sequence mediating unassisted and opportunistic integration into membranes, whereas the other two proteins which must be targeted to particular membrane sites require a signal sequence mediating receptor-dependent integration into specific membranes. We would propose, by analogy, that for some at least of mitochondrial proteins the addressing signal (presequence) and insertion or translocation signals (internal) are distinct. This possibility is developed further in Section VII,D.

A second possibility is that of solubilizing hydrophobic proteins while in the aqueous phase of the cytosol. It is noteworthy in this context that in the case of the proteolipid subunit IX of ATPase from *Neurospora* for which the amino acid sequence of the prepiece is known, the extra sequence is very hydrophilic whereas the mature protein is very hydrophobic (Viebrock *et al.*, 1982). A solubilizing role for the prepiece seems, therefore, very likely. Obviously this is not the only available strategy for solubilization of hydrophobic proteins. An alternative involving formation of a multimeric complex with altered conformation has been demonstrated for the ADP/ATP translocator which lacks a prepeice (Zimmermann and Neupert, 1980). It should be noted in passing, however, that the precursors of both ornithine transcarbamylase and carbamoyl-phosphate synthetase exist in solution as aggregates of high molecular weight even though the native proteins are water soluble (Miura *et al.*, 1981); this suggests that precursor aggregation may be important in other contexts besides solubilization of hydrophobic proteins.

A third possibility could be that the prepiece exists to prevent premature assembly of proteins into oligomeric structures. The majority of mitochondrial proteins exist in the organelles as either homodimers or higher polymers, or as heteromeric complexes. It is very likely that formation of complexes in the cytosol would prevent translocation into the organelles and, in cases where no components synthetized in the mitochondria are required which includes all matrix enzymes, could lead to unwanted enzymic activity in the cytosol with obvious disruption of cellular metabolism. Such a role for the prepiece of the precursor of aspartate aminotransferase seems very likely in view of the fact that in the native dimer the N-terminal segment of each monomer makes contact with the other monomer in such a way as to markedly stabilize the dimeric enzyme (Ford *et al.*, 1980); an N-terminal extension could serve to prevent these contact from forming.

Fourth, the prepiece may function by maintaining the precursor in a nonnative configuration which is required for transmembrane movement or membrane insertion without itself being directly involved in the latter processes; that is, it might facilitate critical interactions between the mitochondrial membrane and internal regions of the polypeptide chain.

Both of the latter two possible roles for the precursors might be considered to be negated by uptake of mature proteins into mitochondria, but this is not necessarily so. We suspect that at the low protein concentrations used in our model system, both aspartate aminotransferase and malate dehydrogenase exist in solution at least partially in a monomeric form. This has been directly demonstrated for malate dehydrogenase (Webster *et al.*, 1979) but not for aspartate aminotransferase. In the latter case, however, cross-linking of the native dimer with dimethyl suberimidate prevents uptake into mitochondria; uptake of chemically modified but non-cross-linked monomer still occurs (K. M. C. O'Donovan and S. Doonan, unpublished results). As for attainment of the correct conformation, we have evidence that a major conformational change occurs when aspartate aminotransferase interacts with the external face of the mitochondrion; this is shown by the fact that the enzyme in solution is extremely resistant to proteases, but becomes very sensitive when bound to mitochondria (Marra *et al.*, 1980a). Webster *et al.* (1980) have shown a similar change in conformation of malate dehydrogenase when bound to liposomes. In this context it is interesting to recall that Clarke (1977) has shown that fumarase can exist in two distinct conformations, one of which binds Triton X-100 and one of which does not.

In summary, then, several roles are possible for the prepiece of the precursor of a mitochondrial protein which do not involve its direct involvement in translocation and which are not mutually exclusive. Moreover these roles are consistent with uptake of mature proteins into mitochondria whereas a direct involvement in translocation is not.

Although not of direct relevance to the subject of this review it should be noted that there have been reports showing that some mitochondrially coded proteins are synthesized as precursors of higher molecular weight than the mature forms (Werner and Bertrand, 1979; Werner *et al.*, 1980; Sevarino and Poyton, 1980; Van't Sant *et al.*, 1981; van den Boogaart *et al.*, 1982). Again, any of the possible roles outlined above could apply for the presequences of these mitochondrial translation products also.

C. POSSIBLE INVOLVEMENT OF THE OTHER REGIONS OF THE STRUCTURE OF IMPORTED PROTEINS

It was argued in Section VII,B that the presequences of precursor proteins may not play a direct role in transmembrane movement or integration. The obvious question then arises as to whether there is any evidence for involvement of other regions of proteins in these processes. In the case of mitochondrial proteins such evidence is sparse. For reasons outlined in Section VI,D no work has been done on chemical modification of precursors of mitochondrial proteins because they have not, up to now, been available in sufficient quantity. An exception to this is for cytochrome c where, as described previously, the chemically prepared apoprotein is functionally equivalent to the precursor. In this context, it has recently been shown (Matsuura et al., 1981) that for the rat liver protein the signal sequence that directs newly synthesized cytochrome c to the inner mitochondrial membrane is located between residues 66 and the C-terminus.

Again, the model system involving import of native proteins provides some further clues on this topic. As described in Section VI,D uptake of both aspartate aminotransferase and malate dehydrogenase into mitochondria is blocked by reaction with sulfhydryl group reagents. In the case of aspartate aminotransferase, reaction of a single cysteine is involved and this residue is probably at position 166 in the polypeptide chain (Marra et al., 1979; Martini et al., 1983). It is difficult to see how modification of a single amino acid residue would block import if the region of the protein modified were not intimately involved in the uptake process. Further, as described in Section VI,D, removal of the N-terminal 27 amino acid residues from the native protein prevents both uptake and binding of the enzyme to mitochondria; again this is indicative of a direct involvement in the uptake process. Less directly, the amino acid sequences of several mitochondrial aspartate aminotransferases are now known and comparison of these structures (Christen et al., 1984) shows that certain limited regions of the enzyme have been extremely highly conserved during evolution in the mitochondrial isoenzyme but not in the cytosolic form; these regions are highly polar and on the surface of the protein. This high degree of conservation of surface regions in the mitochondrial isoenzyme but not in cytosolic form suggests that they are not required for catalytic activity but may play a role in topogenesis of the mitochondrial isoenzyme. Hence the picture that emerges is one of several parts of the protein widely separated in the primary structure being involved in uptake into mitochondria.

The concept of internal sequences acting as signals for transport across membranes is not, of course, a new one. It has been known for some time (Palmiter et al., 1978) that the secretory protein ovalbumin is synthesized without a presequence. Recent work by Meek et al. (1982) has shown that the signal is located near the N-terminus of the native protein, probably between residues 25 to 45. These authors speculate that this sequence forms an amphipathic hair-pin structure which is recognized by the endoplasmic reticulum as the transport signal.

Other examples of internal signal sequences directing protein secretion come from studies of bacteria. For example, a fusion protein consisting of β-galactosidase-preproinsulin is secreted by E. coli cells, secretion being determined by the internal presequence of the preproinsulin (Talmadge et al., 1981).

As well as functioning in protein secretion, internal signal sequences are thought to participate as insertion sequences for proteins that remain integrated into membranes; otherwise it is very difficult to explain the topogenesis of proteins that span a membrane more than once (Blobel, 1980). A particularly well-understood example of this comes from the study of Semliki Forest virus (Garoff *et al.*, 1978, 1980). This and other examples have been recently reviewed by Sabatini *et al.* (1982) and will not be further discussed here except to note that the mechanism referred to above could well play a part in topogenesis of proteins of the inner mitochondrial membrane.

D. NATURE OF THE RECEPTOR SITES

There are clearly two possible candidates for the receptor sites, namely lipid or protein. Blobel (1980) proposes that for all types of translocation or integration the initial event is mediated by protein translocators, whereas for integral proteins internal insertion sequences interact directly with the lipid phase of the membrane. Conversely, in Wickner's "membrane-triggered folding" hypothesis (Wickner, 1980, 1983), the primary interaction is proposed to be with the lipid phase. A refinement of the latter model, termed the helical-hairpin hypothesis, was proposed by Engelman and Steitz (1981). The major difficulty with models involving direct insertion into lipid is the origin of specificity of membrane interaction whereas in Blobel's model this would be dictated by the protein part of the translocation apparatus.

Information on the nature of the receptor site for protein transport into mitocondria is limited, but the results obtained using the model system for aspartate aminotransferase and malate dehydrogenase suggest that proteinaceous receptors exist for these two enzymes and that at least part of the interaction between the proteins and their receptors is mediated by enzyme thiols (Section VI,E); it would be difficult to account for inhibition of protein transport by 2-mercaptoethanol if the receptor were a lipid region of the mitochondrial membrane. Our results further suggest that distinct but interacting receptor sites exist for aspartate aminotransferase and malate dehydrogenase.

Further evidence for proteinaceous receptor sites for import of proteins into mitochondria has come from experiments where mitochondria exposed to proteases have been shown to loose their ability to internalize precursor proteins (see, for example, Gasser *et al.*, 1982b). Similarly, yeast outer membrane vesicles loose their ability to bind precursors when exposed to protease treatment (Riezman *et al.*, 1983b).

There appears to be a conflict between our results and those of Schatz and his colleagues concerning which of the mitochondrial membranes carries the receptor sites for protein uptake. Our observations that uptake of native proteins into intact mitochondria and into mitoplasts are essentially indistinguishable (Section, VI,E) suggests that the receptor sites for internalization of matrix proteins are on the inner membrane. A logical extension of this proposition is that protein uptake might occur at points where the outer membrane and inner membrane fuse transiently to expose the receptor sites in the latter. Riezman *et al.* (1983b) on the other hand have presented evidence for receptor sites in the outer membrane of yeast mitochondria which bind pre-cytochrome b_2 and pre-citrate synthase. Is it possible to reconcile these seemingly conflicting results without recourse to invoking species differences? One possibility would be that, at least for matrix proteins, the complete protein uptake machinery consists of two distinct receptors with different functions, one on the outer membrane and one on the inner membrane.

The function of the receptor on the outer membrane could be in recognition of an addressing signal for proteins destined for import into the organelles, the signal being provided by the N-terminal extensions of precursors. Once recruited onto the mitochondria, import of proteins might then be mediated by the second receptor in the inner membrane, interaction in this case involving multiple sites in the protein to be imported.

Admittedly the evidence for this hypothesis is tenuous, but it has the attraction of resolving some aspects of the conflict between uptake of mature proteins in our model system and biosynthesis of most or all of matrix proteins in precursor form. The role of the prepiece of precursors in this hypothesis would be to direct newly synthesized proteins to the mitochondria rather than to any of the other membranous structures within the cell. This problem does not arise in our model system since no organelles or membranes other than mitochondria are present. Once recruited to the mitochondria, matrix proteins are envisaged as interacting with the inner-membrane receptor via internal regions of the polypeptide chain. Thus uptake of native proteins into mitochondria may occur provided that the binding sites on the inner membrane are accessible in intact mitochondria (perhaps at points where the two membranes are in contact or fuse transiently; see Section VI,E). The observations reported by Riezman *et al.* (1983b) with respect to citrate synthase are consistent with this model. The precursor protein binds to isolated outer membrane vesicles but the mature protein does not; this is as expected since the receptor in the outer membrane is envisaged as being specific for the prepiece. Moreover, the precursor also binds to the inner membrane this binding being competed-out by mature citrate synthase; again this is as expected if an inner-membrane receptor exists which recognizes regions of the structure of the native protein.

Obviously this model is not applicable to proteins of the intermembranal space and probably not to proteins of the inner membrane. In the former case a more extensive role for the prepieces of precursor proteins than simply as an addressing signal has been established (Daum *et al.*, 1982; Ohashi *et al.*, 1982; Reid *et al.*, 1982). Moreover Riezman *et al.* (1983b) have shown that neither the precursor of cytochrome b_2 nor the partially processed form binds to the inner membrane of yeast mitochondria. Hence in this case a single receptor in the outer membrane may be responsible both for precursor binding and subsequent to-pogenesis. A similar situation may obtain with inner membrane proteins.

E. ENERGETICS OF THE PROCESS

Initial reports (Nelson and Schatz, 1979) that integration of proteins into the inner mitochondrial membrane is directly dependent on ATP have been shown to be in error. It is now clear that in most cases (exceptions being the outer-membrane protein porin and cytochrome c) protein transport into mitochondria requires a transmembrane electrochemical potential. The evidence for this is summarized in Section VI,B as are the reasons for maintaining that it is the proton component of such a gradient that is essential, at least in the case of proteins destined for the matrix space.

The mechanism by which a proton gradient is coupled to transmembrane movement is by no means clear. Robillard and Konings (1982) have presented an interesting general hypothesis for solute transport across membranes involving dithiol–disulfide interchanges. They propose that affinities of substrate binding sites are regulated by protein dithiol and disulfide groups at different depths in membranes and that the coupled redox states of these groups are regulated by a transmembrane electrical potential or pH gradient. In the case of protein transport into mitochondria, therefore, a high affinity dithiol-containing receptor protein on the cytoplasmic side of the membrane might bind the protein to be imported, transport across the membrane being accompanied by oxidation of the dithiol to a disulfide and concomitant reduction of the disulfide group on the inner face of the membrane system. However, apart from the fact that analogies between transport of low-molecular-weight solutes and of macromolecules are obviously artificial, a mechanism such as that described above is rendered improbable by the fact that treatment of mitochondria with mersalyl does not impair their ability to import proteins (Passarella et al., 1980).

An alternative view of the role of a transmembrane pH gradient is one solely in terms of energetics. Proteins on the cytoplasmic side of the membrane system would be more highly protonated than after translocation into the alkaline matrix. Loss of protons from the proteins on translocation could confer an overall negative free energy change on the total process. It is usually assumed that unidirectionality of uptake of proteins into mitochondria is a result of a negative free energy change associated with the conversion of the precursor form into the mature form but there is no reasons a priori to assume that this process would be spontaneous in the absence of changes of protonation state. Indeed the observations of Swizinski and Neupert (1983) that mitochondria in which the processing protease is inactivated still import and accumulate precursor proteins (see Section VI,A) provide direct evidence that processing is not the driving force for uptake.

F. Summary

There is still much that is obscure concerning the transport of proteins into or through the mitochondrial membrane systems. In addition, as pointed out previously, it is unlikely that the details of the process are the same for proteins destined for different compartments of the organelle. A brief summary of the process for matrix proteins might be as follows:

1. The proteins are synthesized on free polysomes as precursors of higher molecular weight than the native forms.

2. These precursors are liberated into the cell cytosol and subsequently translocated into the mitochondria. This timing might be different in yeast under some circumstances, synthesis being completed in association with the mitochondria.

3. The precursors interact with a receptor in the outer mitochondrial membrane interaction being mediated by the presequences of the precursors. The presequences therefore act as addressing signals as well as possibly playing a role in one or all of (a) solubilization of precursors, (b) prevention of premature assembly into multimeric structures, or (c) maintenance of nonnative configurations required for transport.

4. Interaction occurs with a second receptor, this time in the inner membrane of the mitochondria, interaction being with multiple sites in the polypeptide chain.

5. Transport across the inner membrane then occurs, this transport depending on a transmembrane electrochemical gradient of which the proton component is the essential part.

6. Transport is accompanied or followed by proteolysis of the prepiece, and formation of the native structure.

While steps 1 and 2 of this sequence can be considered well established, the remaining steps are still poorly understood or purely hypothetical. Nevertheless, this sequence of events is consistent with known facts about the process and provides a framework for future investigations.

References

Ades, I. Z. (1982). *Mol. Cell. Biochem.* **43**, 113–127.
Ades, I. Z., and Butow, R. A. (1980a). *J. Biol. Chem.* **255**, 9918–9924.
Ades, I. Z., and Butow, R. A. (1980b). *J. Biol. Chem.* **255**, 9925–9935.
Ades, I. Z., and Harpe, K. G. (1981). *J. Biol. Chem.* **256**, 9329–9333.
Alam, T., Finkelstein, D., and Srere, P. A. (1982). *J. Biol. Chem.* **257**, 11181–11185.
Anderson, S., Bankier, A. T., Barrel, B. G., de Bruijn, M. H. L., Coulson, A. R., Drouin, J.

Eperon, I. C., Nierlich, D. P., Roe, B. A., Sanger, F., Schreier, P. H., Smith, A. J. H., Standen, R., and Young, I. G. (1981). *Nature (London)* **290,** 457–465.

Anderson, S., de Bruijn, M. H. L., Coulson, A. R., Eperon, I. C., Sanger, F., and Young, I. G. (1982). *J. Mol. Biol.* **156,** 683–717.

Anderson, D. J., Mostov, K. E., and Blobel, G. (1983). *Proc. Natl. Acad. Sci. U.S.A.* **80,** 7249–7253.

Argan, C., Lusty, C. J., and Shore, G. C. (1983). *J. Biol. Chem.* **258,** 6667–6670.

Autor, A. P. (1982). *J. Biol. Chem.* **257,** 2713–2718.

Aziz, L. E., Chien, S.-M., Patel, H. V., and Freeman, K. B. (1981). *FEBS Lett.* **133,** 127–129.

Bibb, M. J., Van Etten, R. A., Wright, C. T., Walberg, M. W., and Clayton, D. A. (1981). *Cell* **26,** 167–180.

Bingham, R. N., and Campbell, P. N. (1972). *Biochem. J.* **126,** 211–215.

Blobel, G. (1980). *Proc. Natl. Acad. Sci. U.S.A.* **77,** 1496–1500.

Blobel, G., and Dobberstein, B. (1975). *J. Cell Biol.* **67,** 852–862.

Böhni, P., Gasser, S., Leaver, C., and Schatz, G. (1980). *In*"The Organization and Expression of the Mitochondrial Genome" (A. M. Kroon and C. Saccone, eds.), pp. 423–433. Elsevier, Amsterdam.

Bossa, F., Barra, D., Martini, F., Schininà, E., Doonan, S., and O'Donovan, K. M. C. (1981). *Comp. Biochem. Physiol.* **69B,** 753–760.

Chien, S.-M., and Freeman, K. B. (1983). *Fed. Proc. Fed. Am. Soc. Exp. Biol.* **42,** 2125.

Christen, P., Graf-Hausner, U., Bossa, F., and Doonan, S. (1984). *In* "Transaminases" (P. Christen and D. E. Metzler, eds.). Wiley, New York, in press.

Chua, N.-H., and Schmidt, G. W. (1979). *J. Cell Biol.* **81,** 461–483.

Clarke, S. (1977). *Biochem. Biophys. Res. Commun.* **79,** 46–52.

Conboy, J. G., Kalousek, F., and Rosenberg, L. E. (1979). *Proc. Natl. Acad. Sci. U.S.A.* **76,** 5724–5727.

Côté, C., Solioz, M., and Schatz, G. (1979). *J. Biol. Chem.* **254,** 1437–1439.

Cumsky, M. G.. McEwen, J. E., Ko, C., and Poyton, R. O. (1983). *J. Biol. Chem.* **258,** 13418–13421.

Daum, G., Gasser, S. M., and Schatz, G. (1982). *J. Biol. Chem.* **257,** 13075–13080.

De Jong, L., Holtrop, M., and Kroon, A. M. (1980). *Biochim. Biophys. Acta* **608,** 32–38.

Dujardin, G., Jacq, C., and Slonimsky, P. P. (1982). *Nature (London)* **298,** 628–632.

Edwards, Y. H., and Hopkinson, D. A. (1979). *Ann. Hum. Genet.* **42,** 303–313.

Eiermann, W., Aquila, H., and Klingenberg, M. (1977). *FEBS Lett.* **74,** 209–214.

Engelman, D. M., and Steitz, T. A. (1981). *Cell* **23,** 411–422.

Felipo, V., Miralles, V., Knecht, E., Hernandez-Yago, J., and Grisolia, S. (1983). *Eur. J. Biochem.* **133,** 641–644.

Flurkey, W., Kim, Y. S., and Kolattukudy, P. E. (1982). *Biochem. Biophys. Res. Commun.* **106,** 1346–1352.

Ford, G. C., Eichele, G., and Jansonius, J. N. (1980). *Proc. Natl. Acad. Sci. U.S.A.* **77,** 2559–2563.

Freeman, K. B., Chien, S.-M., Litchfield, D., and Patel, H. V. (1983). *FEBS Lett.* **158,** 325–330.

Freitag, H., Janes, M., and Neupert, W. (1982). *Eur. J. Biochem.* **126,** 197–202.

Furuya, E., Yoshida, Y., and Tagawa, K. (1979). *J. Biochem. Tokyo* **85,** 1157–1163.

Garoff, H., Simons, K., and Dobberstein, B. (1978). *J. Mol. Biol.* **124,** 587–600.

Garoff, H., Frischauf, A.-M., Simons, K., Lehrach, H., and Delius, H. (1980). *Nature (London)* **288,** 236–241.

Gasser, S. M., and Schatz, G. (1983). *J. Biol. Chem.* **258,** 3427–3430.

Gasser, S. M., Ohashi, A., Daum, G., Böhni, P. C., Gibson, J., Reid, G. A., Yonetani, T., and Schatz, G. (1982a). *Proc. Natl. Acad. Sci. U.S.A.* **79,** 267–271.

Gasser, S. M., Daum, G., and Schatz, G. (1982b). *J. Biol. Chem.* **257**, 13034–13041.

Gilmore, R., Blobel, G., and Walter, P. (1982a). *J. Cell Biol.* **95**, 463–469.

Gilmore, R., Walter, P., and Blobel, G. (1982b). *J. Cell Biol.* **95**, 470–477.

Godinot, C., and Lardy, H. A. (1973). *Biochemistry* **12**, 2051–2061.

Gonzalez-Cadavid, N. F., and de Cordova, C. S. (1974). *Biochem. J.* **140**, 157–167.

Goto, Y., Ohki, Y., Shimizu, J., and Shukuya, R. (1981). *Biochim. Biophys. Acta* **657**, 383–389.

Hallermayer, G., Zimmermann, R., and Neupert, W. (1977). *Eur. J. Biochem.* **81**, 523–532.

Hampsey, D. M., Lewin, A. S., and Kohlaw, G. B. (1983). *Proc. Natl. Acad. Sci. U.S.A.* **80**, 1270–1274.

Harmey, M. A., and Neupert, W. (1979). *FEBS Lett.* **108**, 385–389.

Harmey, M. A., Hallermayer, G., Korb, H., and Neupert, W. (1977). *Eur. J. Biochem.* **81**, 533–544.

Hase, T., Riezman, H., Suda, K., and Schatz, G. (1983). *EMBO J.* **2**, 2169–2172.

Hattori, T., Iwasaki, Y., Sakajo, S., and Asahi, T. (1983). *Biochem. Biophys. Res. Commun.* **113**, 235–240.

Hay, R., Böhni, P., and Gasser, S. (1984). *Biochim. Biophys. Acta* **779**, 65–87.

Hayashi, H., Katunuma, N., Chiku, K., Endo, Y., and Natori. Y. (1981). *J. Biochem. Tokyo* **90**, 1229–1232.

Hennig, G. B., and Neupert, W. (1981). *Eur. J. Biochem.* **121**, 203–217.

Hennig, B., Koehler, H., and Neupert, W. (1983). *Proc. Natl. Acad. Sci. U.S.A.* **80**, 4963–4967.

Hensgens, L. A. M., Grivell, L. A., Borst, P., and Bos, J. L. (1979). *Proc. Natl. Acad. Sci. U.S.A.* **76**, 1663–1667.

Hod, Y., Utter, M. F., and Hanson, R. W. (1982). *J. Biol. Chem.* **257**, 13787–13794.

Hopper, A. K., Furukawa, A. H., Pham, H. D., and Martin, N. C. (1982). *Cell* **28**, 543–550.

Horwich, A. L., Kraus, J. P., Williams, K., Kalousek, F., Konigsberg, W., and Rosenberg, L. E. (1983). *Proc. Natl. Acad. Sci. U.S.A.* **80**, 4258–4262.

Hubert, P., Crémel, G., Rendon, A., Sacko, B., and Waksman, A. (1979). *Biochemistry* **18**, 3119–3126.

Jaussi, R., Sonderegger, P., Flückiger, J., and Christen, P. (1982). *J. Biol. Chem.* **257**, 13334–13340.

Kamisaki, Y., Sakakibara, R., Horio, Y., and Wada, H. (1982). *Biochem. Int.* **4**, 289–296.

Kaput, J., Goltz, S., and Blobel, G. (1982). *J. Biol. Chem.* **257**, 15054–15058.

Kellems, R. E., and Butow, R. A. (1972). *J. Biol. Chem.* **247**, 8043–8052.

Kellems, R. E., Allison, V. F., and Butow, R. A. (1975). *J. Cell Biol.* **65**, 1–14.

Kikuchi, G., and Hayashi, N. (1981). *Mol. Cell. Biochem.* **37**, 27–41.

Kim, Y. S., and Kolattukudy, P. E. (1978a). *Arch. Biochem. Biophys.* **190**, 234–246.

Kim, Y. S., and Kolattukudy, P. E. (1978b). *Biochim. Biophys. Acta* **531**, 187–196.

Kitagawa, Y., Murakami, A., and Sugimoto, E. (1984) *FEBS Lett.* **165**, 133–137.

Klingenberg, M. (1970). *Essays Biochem.* **6**, 117–159.

Korb, H., and Neupert, W. (1978). *Eur. J. Biochem.* **91**, 609–620.

Kraus, J. P., Conboy, J. G., and Rosenberg, L. E. (1981). *J. Biol. Chem.* **256**, 10739–10742.

Kreil, G. (1981). *Annu. Rev. Biochem.* **50**, 317–348.

Lewin, A. S., Gregor, I., Mason, T. L., Nelson, N., and Schatz, G. (1980). *Proc. Natl. Acad. Sci. U.S.A.* **77**, 3998–4002.

Lindén, M., Gellerfors, P., and Nelson, B. D. (1982). *Biochem. J.* **208**, 77–82.

Lustig, A., Levens, D., and Rabinowitz, M. (1982). *J. Biol. Chem.* **257**, 5800–5808.

Maccecchini, M.-L., Rudin, Y., Blobel, G., and Schatz, G. (1979a). *Proc. Natl. Acad. Sci. U.S.A.* **73**, 343–347.

Maccecchini, M.-L., Rudin, Y., and Schatz, G. (1979b). *J. Biol. Chem.* **254**, 7468–7471.

Macino, G., and Tzagoloff, A. (1979). *J. Biol. Chem.* **254**, 4617–4623.

McGraw, P., and Tzagoloff, A. (1983). *J. Biol. Chem.* **258**, 9459–9468.

Mannella, C. A. (1982). *J. Cell Biol.* **94**, 680–687.

Marra, E., Doonan, S., Saccone, C., and Quagliariello, E. (1977a). *In* "Bioenergetics of Membranes" (L. Packer, G. C. Papageorgiou, and A. Trebst, eds.), pp. 415–426. Elsevier, Amsterdam.

Marra, E., Doonan, S., Saccone, C., and Quagliariello, E. (1977b). *Biochem. J.* **164**, 685–691.

Marra, E., Doonan, S., Saccone, C., and Quagliariello, E. (1978). *Eur. J. Biochem.* **83**, 427–435.

Marra, E., Passarella, S., Doonan, S., Saccone, C., and Quagliariello, E. (1979). *Arch. Biochem. Biophys.* **195**, 269–279.

Marra, E., Passarella, S., Doonan, S., Quagliariello, E., and Saccone, C. (1980a). *FEBS Lett.* **122**, 33–36.

Marra, E., Passarella, S., Doonan, S., Quagliariello, E., and Saccone, C. (1980b). *In* "The Organization and Expression of the Mitochondrial Genome" (A. M. Kroon and C. Saccone, eds.), pp. 435–438. Elsevier, Amsterdam.

Marra, E., Passarella, S., Doonan, S., Casamassima, E., Quagliariello, E., and Saccone, C. (1982). *Ital. J. Biochem.* **31**, 289–290.

Marra, E., Passarella, S., Doonan, S., Serafino, M. R., Saccone, C., and Quagliariello, E. (1982b). *Bull. Mol. Biol. Med.* **7**, 97–104.

Martini, F., Angelaccio, S., Barra, D., Doonan, S., and Bossa, F. (1983). *Comp. Biochem. Physiol.,* **76B**, 483–487.

Matsuura, S., Arpin, M., Hannum, C., Margoliash, E., Sabatini, D. D., and Morimoto, T. (1981). *Proc. Natl. Acad. Sci. U.S.A.* **78**, 4368–4372.

McAda, P. C., and Douglas, M. G. (1982). *J. Biol. Chem.* **257**, 3177–3182.

Meek, R. L., Walsh, K. A., and Palmiter, R. D. (1982). *J. Biol. Chem.* **257**, 12245–12251.

Michel, R., Wachter, E., and Sebald, W. (1979). *FEBS Lett.* **101**, 373–376.

Mihara, K., and Blobel, G. (1980). *Proc. Natl. Acad. Sci. U.S.A.* **77**, 4160–4164.

Mihara, K., Omura, T., Harano, T., Brenner, S., Fleischer, S., Rajagopalan, K. V., and Blobel, G. (1982a). *J. Biol. Chem.* **257**, 3355–3358.

Mihara, K., Blobel, G., and Sato, R. (1982b). *Proc. Natl. Acad. Sci. U.S.A.* **79**, 7102–7106.

Miralles, V., Felipo, V., Hernández-Yago, J., and Crisolía, S. (1982). *Biochem. Biophys. Res. Commun.* **107**, 1028–1036.

Mitchell, P. (1968). "Chemiosmotic Coupling and Energy Transduction." Glynn Research, Bodmin.

Miura, S, Mori, M., Amaya, Y., Tatibana, M., and Cohen, P. P. (1981). *Biochem. Int.* **2**, 305–312.

Miura, S., Mori, M., Amaya, Y., and Tatibana, M. (1982a). *Eur. J. Biochem.* **122**, 641–647.

Miura, S., Mori, M., Morita, T., and Tatibana, M. (1982b). *Biochem. Int.* **4**, 201–208.

Miura, S., Mori, M., and Tatibana, M. (1983). *J. Biol. Chem.* **258**, 6671–6674.

Mori, M., Miura, S., Tatibana, M., and Cohen, P. P. (1979a). *Proc. Natl. Acad. Sci. U.S.A.* **76**, 5071–5075.

Mori, M., Morris, S. M., and Cohen, P. P. (1979b). *Proc. Natl. Acad. Sci. U.S.A.* **76**, 3179–3183.

Mori, M., Miura, S., Tatibana, S., and Cohen, P. P. (1980). *J. Biochem. Tokyo* **88**, 1829–1836.

Mori, M., Morita, T., Ikeda, F., Amaya, Y., Tatibana, M., and Cohen, P. P. (1981a). *Proc. Natl. Acad. Sci. U.S.A.* **78**, 6056–6060.

Mori, M., Morita, T., Miura, S., and Tatibana, M. (1981b). *J. Biol. Chem.* **256**, 8263–8266.

Mori, M., Miura, S., Morita, T., Takiguchi, M., and Tatibana, M. (1982). *Mol. Cell. Biochem.* **49**, 97–111.

Morita, T., Mori, M., Ikeda, F., and Tatibana, M. (1982). *J. Biol. Chem.* **257**, 10547–10550.

Mueckler, M. M., Himeno, M., and Pitot, H. C. (1982). *J. Biol. Chem.* **257**, 7178–7180.

Nagata, S., Tsunetsugu-Yokata, Y., Naito, A., and Kaziro, Y. (1983). *Proc. Natl. Acad. Sci. U.S.A.* **80**, 6192–6196.

Nakakuki, M., Yamauchi, K., Hayashi, N., and Kikuchi, G. (1980). *J. Biol. Chem.* **255**, 1738–1745.

Nelson, N., and Schatz, G. (1979). *Proc. Natl. Acad. Sci. U.S.A.* **76**, 4365–4369.

Neupert, W., and Schatz, G. (1981). *Trends Biochem. Sci.* **6,** 1–4.

Oda, T., Ichiyama, A., Miura, S., Mori, M., and Tatibana, M. (1981). *Biochem. Biophys. Res. Commun.* **102,** 568–573.

O'Donovan, K. M. C., Doonan, S., Marra, E., and Passarella, S. (1984). *Biochem. Soc. Trans.* **12,** 444–445.

O'Hare, M. C., and Doonan, S. (1984). *Biochem. Soc. Trans.* **12,** 443–444.

Ohashi, A., Gibson, J., Gregor, I., and Schatz, G. (1982). *J. Biol. Chem.* **257,** 13042–13047.

Ono, H., and Ito, A. (1982). *Biochem. Biophys. Res. Commun.* **107,** 258–264.

Ono, H., and Ito, A., and Omura, T. (1982). *J. Biochem. Tokyo* **91,** 107–116.

Palmieri, F., and Quagliariello, E. (1969). *Eur. J. Biochem.* **8,** 473–481.

Palmiter, R. D., Gagnon, J., and Walsh, K. A. (1978). *Proc. Natl. Acad. Sci. U.S.A.* **75,** 94–98.

Passarella, S., Marra, E., Doonan, S., and Quagliariello, E. (1980). *Biochem. J.* **192,** 649–658.

Passarella, S., Marra, E., Doonan, S., Languino, L. R., Saccone, C., and Quagliariello, E. (1982a). *Biochem. J.* **202,** 353–362.

Passarella, S., Marra, E., Doonan, S., Perlino, E., Saccone, C., and Quagliariello, E. (1982b). *Ital. J. Biochem.* **31,** 291–292.

Passarella, S., Marra, E., Doonan, S., and Quagliariello, E. (1983). *Biochem. J.* **210,** 207–214.

Pepe, G., Holtrop, M., Gadaleta, G., Kroon, A. M., Cantatore, P., Gallerani, R., De Benedetto, C., Quagliariello, C., Sbisà, E., and Saccone, C. (1983). *Biochem. Int.* **6,** 553–563.

Perryman, M. B., Strauss, A. W., Olson, J., and Roberts, R. (1983). *Biochem. Biophys. Res. Commun.* **110,** 967–972.

Petrilli, P., Pucci, P., Garzillo, A. M., Sannia, G., and Marino, G. (1981). *Mol. Cell. Biochem.* **35,** 121–128.

Póyton, R. O., and McKemmie, E. (1979a). *J. Biol. Chem.* **254,** 6763–6771.

Poyton, R. O., and McKemmie, E. (1979b). *J. Biol. Chem.* **254,** 6772–6780.

Raymond, Y., and Shore, G. C. (1981). *J. Biol. Chem.* **256,** 2087–2090.

Reid, G. A., and Schatz, G. (1982a). *J. Biol. Chem.* **257,** 13056–13061.

Reid, G. A., and Schatz, G. (1982b). *J. Biol. Chem.* **257,** 13062–13067.

Reid, G. A., Yonetani, T., and Schatz, G. (1982). *J. Biol. Chem.* **257,** 13068–13074.

Riezman, H., Hay, R., Gasser, S., Daum, G., Schneider, G., Witte, C., and Schatz, G. (1983a). *EMBO J.* **2,** 1105–1111.

Riezman, H., Hay, R., Witte, C., Nelson, N., and Schatz, G. (1983b). *EMBO J.* **2,** 1113–1118.

Riezman, H., Hase, T., van Loon, A. P. G. M., Grivell, L. A., Suda, K., and Schatz, G. (1983c). *EMBO J.* **2,** 2161–2168.

Robillard, G. T., and Konings, W. N. (1982). *Eur. J. Biochem.* **127,** 597–604.

Rosamond, J. (1982). *Biochem. J.* **202,** 1–8.

Sabatini, D. D., Kreibich, G., Morimoto, T., and Adesnik, M. (1982). *J. Cell Biol.* **92,** 1–22.

Sakakibara, R., Huynh, Q. K., Nishida, Y., Watanabe, T., and Wada, H. (1980). *Biochem. Biophys. Res. Commun.* **95,** 1781–1788.

Sakakibara, R., Kamisaki, Y., and Wada, N. (1981). *Biochem. Biophys. Res. Commun.* **102,** 235–242.

Saltzgaber-Muller, J., Kunapuli, S. P., and Douglas, M. G. (1983). *J. Biol. Chem.* **258,** 11465–11470.

Sandmeier, E., and Christen, P. (1980). *J. Biol. Chem.* **255,** 10284–10289.

Sannia, G., Abrescia, P., Colombo, M., Giardina, P., and Marino, G. (1982). *Biochem. Biophys. Res. Commun.* **105,** 444–449.

Sannia, G., Colombo, M., Palomba, R., and Marino, G. (1983). *Biochem. Int.* **6,** 731–736.

Schatz, G. (1979). *FEBS Lett.* **103,** 203–211.

Schatz, G., and Butow, R. A. (1983). *Cell* **32,** 316–318.

Scherer, G., Schmid, W., Strange, C. H., Röwekamp, W., and Schütz, G. (1982). *Proc. Natl. Acad. Sci. U.S.A.* **79,** 7205–7208.

Schleyer, M., Schmidt, B., and Neupert, W. (1982). *Eur. J. Biochem.* **125**, 109–116.

Schmelzer, E., and Heinrich, P. C. (1980). *J. Biol. Chem.* **255**, 7503–7506.

Schmidt, B., Hennig, B., Zimmermann, R., and Neupert, W. (1983a). *J. Cell Biol.* **96**, 248–255.

Schmidt, B., Hennig, B., Köhler, H., and Neupert, W. (1983b). *J. Biol. Chem.* **258**, 4687–4689.

Sebald, W. (1977). *Biochim. Biophys. Acta* **463**, 1–27.

Sevarino, K. A., and Poyton, R. O. (1980). *Proc. Natl. Acad. Sci. U.S.A.* **77**, 142–146.

Shore, G. C., and Tata, J. R. (1977a). *J. Cell Biol.* **72**, 714–725.

Shore, G. C., and Tata, J. R. (1977b). *J. Cell Biol.* **72**, 726–743.

Shore, G. C., and Tata, J. R. (1977c). *Biochim. Biophys. Acta* **472**, 197–236.

Shore, G. C., Carignan, P., and Raymond, Y. (1979). *J. Biol. Chem.* **254**, 3141–3144.

Shore, G. C., Power, F., Bendayan, M., and Carignan, P. (1981). *J. Biol. Chem.* **256**, 8761–8766.

Sonderegger, P., Jaussi, R., and Christen, P. (1980). *Biochem. Biophys. Res. Commun.* **94**, 1256–1260.

Sonderegger, P., Jaussi, R., Christen, P., and Gehring, H. (1982). *J. Biol. Chem.* **257**, 3339–3345.

Strauss, A. W., and Boime, I. (1982). *Crit. Rev. Biochem.* **13**, 205–235.

Suissa, M., and Schatz, G. (1982). *J. Biol. Chem.* **257**, 13048–13055.

Suominen, I., and Mäntsälä, P. (1983). *Int. J. Biochem.* **15**, 591–601.

Takiguchi, M., Miura, S., Mori, M., and Tatibana, H. (1983). *Comp. Biochem. Physiol.* **75B**, 227–231.

Talmadge, K., Brosius, J., and Gilbert, W. (1981). *Nature (London)* **294**, 176–178.

Teintze, M., Slaughter, M., Weiss, H., and Neupert, W. (1982). *J. Biol. Chem.* **257**, 10364–10371.

Tzagoloff, A. (1982). "Mitochondria." Plenum, New York.

Tzagoloff, A., Macino, G., and Sebald, W. (1979). *Annu. Rev. Biochem.* **48**, 419–441.

Van den Boogaart, P., Van Dijk, S., and Agsteribbe, E. (1982). *FEBS Lett.* **147**, 97–100.

van Loon, A. P. G. M., de Groot, R. J., van Eyk, E., van der Horst, G. T. J., and Grivell, L. A. (1982). *Gene* **20**, 323–337.

Van't Sant, P., Mak, J. F. C., and Kroon, A. M. (1981). *Eur. J. Biochem.* **121**, 21–26.

Viebrock, A., Perz, A., and Sebald, W. (1982). *EMBO J.* **1**, 565–571.

Waksman, A., and Rendon, A. (1974). *Biochimie* **56**, 907–924.

Waksman, A., Rendon, A., Crémel, G., Pellicone, C., and Goubault de Brugière, J.-F. (1977). *Biochemistry* **16**, 4703–4707.

Waksman, A., Hubert, P., Crémel, G., Rendon, A., and Burgun, C. (1980). *Biochim. Biophys. Acta* **604**, 249–296.

Wallace, D. C. (1982). *Microbiol. Rev.* **46**, 208–240.

Watanabe, K., and Kubo, S. (1982). *Eur. J. Biochem.* **123**, 587–592.

Webster, K. A., Patel, H. V., Freeman, K. B., and Papahadjopoulos, D. (1979). *Biochem. J.* **178**, 147–158.

Webster, K. A., Freeman, K. B., and Ohki, S. (1980). *Biochem. J.* **186**, 227–233.

Werner, S., and Bertrand, H. (1979). *Eur. J. Biochem.* **99**, 463–470.

Werner, S., Machleidt, W., Bertrand, H., and Wild, G. (1980). *In* "The Organization and Expression of the Mitochondrial Genome" (A. M. Kroon and C. Saccone, eds.), pp. 399–411. Elsevier, Amsterdam.

Wickner, W. (1979). *Annu. Rev. Biochem.* **48**, 23–45.

Wickner, W. (1980). *Science* **210**, 861–868.

Wickner, W. (1983). *Trends Biochem. Sci.* **8**, 90–94.

Yamauchi, K., Hayashi, N., and Kikuchi, G. (1980). *J. Biol. Chem.* **255**, 1746–1751.

Zimmermann, R., and Neupert, W. (1980). *Eur. J. Biochem.* **109**, 217–229.

Zimmermann, R., Paluch, V., Sprinzl, M., and Neupert, W. (1979). *Eur. J. Biochem.* **99**, 247–252.

Zimmermann, R., Hennig, B., and Neupert, W. (1981). *Eur. J. Biochem.* **116**, 455–460.

Zwizinski, C., and Neupert, W. (1983). *J. Biol. Chem.* **258**, 13340–13346.

Zwizinski, C., Schleyer, M., and Neupert, W. (1983). *J. Biol. Chem.* **258**, 4071–4074.

Index

Contents of Recent Volumes